Designing Professional Websites with Odoo Website Builder

Create and customize state-of-the-art websites and
e-commerce apps for your modern business needs

Sainu Nannat

BIRMINGHAM—MUMBAI

Designing Professional Websites with Odoo Website Builder

Group Product Manager: Aaron Lazar
Publishing Product Manager: Denim Pinto
Senior Editor: Rohit Singh
Content Development Editor: Kinnari Chohan
Technical Editor: Gaurav Gala
Copy Editor: Safis Editing
Project Coordinator: Francy Puthiry
Proofreader: Safis Editing
Indexer: Manju Arasan
Production Designer: Jyoti Chauhan

First published: May 2021

Production reference: 2080721

Published by Packt Publishing Ltd.
Livery Place
35 Livery Street
Birmingham
B3 2PB, UK.

ISBN 978-1-80107-812-2

www.packt.com

*To my family, especially my parents, for supporting me all these years.
I would like to dedicate this book to all my friends and colleagues, along
with my subordinates and supporters, for helping me write this book.*

– Sainu Nannat

Contributors

About the author

Sainu Nannat is an Indian entrepreneur, researcher, business analyst, and investor. He is the founder and CEO of Cybrosys Technologies and Blockchain Expert London and the co-founder and CTO of Luvia Digital Ltd. For the past 13 years, he has worked in the area of ERP implementation, gaining expertise in processing and analyzing existing business strategies, such as strategic planning and streamlining operating procedures. Throughout his career, he has been able to analyze and incorporate a wide range of technologies, including Odoo ERP, blockchain, and the Internet of Things.

I want to thank the people who have been close to me and supported me.

About the reviewer

Ila Rana possesses Odoo V14 and Odoo V12 Functional Certificates. She works as an Odoo techno-functional consultant. Her career started with the Odoo ERP product framework in 2011. She has gained experience in various job roles, such as Odoo developer, implementer, consultant, trainer, and project lead. She has worked on more than 60 Odoo projects and has customized localization (accounting, and HR/payroll) for various regions (UAE, Saudi Arabia, Singapore, and Hong Kong). She also took the opportunity to expand her knowledge (development skills and solution architecture design) for different production-line businesses, such as casting MRP, furniture, school uniform, cable wire, and cosmetic production. She has also provided technical and functional training to Odoo clients and developers.

Credits

Dr. V. Kabeer is presently working as Assistant Professor and Head of the Department of Computer Science, Farook College, Kozhikode, Kerala, India. He also leads the Digital Wing of the College. His team developed complete software for the automation of the college's entire processes in Odoo. As a Python programmer, he has coded applications for AI and deep learning. He has published about 30 research papers and given more than 60 invited talks in his areas of research.

Jafar Shareef is a project manager and software developer at Cybrosys Technologies. He is a functionally & technically trained Odoo specialist with a thorough understanding of the Odoo system. He began his career as a developer at Cybrosys Technologies and has spent 10+ years working on R&D and the development of various applications. During this time, he has implemented ERPs for various organizations. He has also delivered various training programs for developers in Dot Net and Odoo Framework.

Aneez K is a consultant at Cybrosys Technologies. He is an Odoo functional expert, who assists in analyzing and translating the business processes of clients into Odoo projects. He has also delivered Odoo functional training for many clients and developers.

Evin Davis has been a Content writer for Cybrosys Technologies. He is a trained content writer for Odoo with a good caliber of creating manuals and preparing well-structured drafts in Odoo.

Table of Contents

Section 2: Website Builder in Depth

3
An Introduction to Blocks

4
Design Using Features Blocks

5

Designing a Website using Dynamic Content

6

Inner Content Block Tools

Section 3: Practical Tools

10

A Discussion Forum for Your Clients

11

Tracking Your Website with Odoo

12
Drafting a Contact Page

13
Communicating with Live Chat

Assessments

Other Book You May Enjoy

Index

Preface

The Odoo website builder is an operational tool within the Odoo platform that allows you to design, develop, and manage websites. This book introduces and explains all the features of the Odoo website builder that will help you to be more productive while creating websites.

The book starts with an overview of the Odoo website builder, its functionalities, and the tools it offers. Using descriptive illustrations and practical examples, you'll gain detailed insights into the block operations of the website builder and learn how to work with structure blocks, features blocks, and dynamic content blocks in Odoo.

After that, you'll get an overview of other practical tools, focusing on the functional aspects of the Odoo website builder and looking at the details of the blocks and the operations involving them. As you advance, you'll discover how HTML, CSS, and JavaScript editors can be used in Odoo website builder applications for customization. This Odoo book will take you through the different aspects of website building and show you how e-commerce websites can be designed and developed using website builder applications. You'll create, manage, and run a discussion forum in Odoo using the website builder, and apply your knowledge to add a live chat tool that can be incorporated into your website.

By the end of this book, you'll have gained a solid understanding of the Odoo website builder application and will be able to leverage its features to develop your own websites.

Who this book is for

This book is for Odoo users, functional consultants, techno-functional consultants, web designers, and anyone looking to create impressive websites. Odoo developers will also find the book useful for building their website for the end user. Basic functional knowledge of Odoo is all that you need to get started with this book.

What this book covers

Chapter 1, Introduction to Odoo and Its Website Builder, gives you an overview of the Odoo platform and its website builder tools. In this chapter, you will get a basic understanding of the functionalities of the Odoo website builder and how they help you build websites.

Chapter 2, The Website Builder in Action, provides you with an understanding of the basic tools required to develop a stunning website. More specifically, you will learn how to create a new page on a website and edit the color and content of your website. Along with that, you will learn how to edit themes from the backend without coding knowledge.

Chapter 3, Introduction to Blocks – Structure Blocks, gives you a detailed overview of how to design, draft, and build a website using structure blocks.

Chapter 4, Design Using Features Blocks, explores how to design a website using features blocks in the Odoo website builder.

Chapter 5, Designing a Website using Dynamic Content, covers website design using dynamic content type blocks in the Odoo website builder.

Chapter 6, Inner Content Block Tools, explores website design using the inner content blocks of the Odoo website builder, showcased with the help of detailed examples.

Chapter 7, Using the HTML/CSS/JavaScript Editors, provides technical insights as to where the HTML/CSS/JavaScript editor tools can be employed to help you build an attractive website.

Chapter 8, Creating Your Own Blog Pages, covers a detailed overview of using Odoo for blogs. This will allow you to get started in no time on creating and designing a blog with the Odoo website builder.

Chapter 9, Go Live with Your E-Commerce Website, explores how e-commerce websites can be designed and developed in Odoo using the website builder applications.

Chapter 10, A Discussion Forum for Your Clients, gives you an overview of how to create, manage, and run a discussion forum in Odoo efficiently and effectively using the website builder. You will cover all aspects of the functioning of a discussion forum of a website.

Chapter 11, Tracking Your Website with Odoo, explains how you can track your website visitors using the Odoo website builder with various tools.

Chapter 12, Drafting a Contact Page, gives you an insight into how you can design and create a contact page on your website using the Odoo website builder tools. In addition, how to create a contact form for your web page will be discussed.

Chapter 13, Communicating with Live Chat, provides you with a functional description of the live chat tool of the Odoo website builder. In addition, the chapter covers adding, managing, and assigning employees to live chat operations.

To get the most out of this book

Although you might be familiar with the website building applications in Odoo, if you are a beginner in the field, you might need an understanding of Odoo and its website builder module. Moreover, a system with Odoo installed will be required to run the platform and design your website using the website builder.

Odoo releases a new version almost every year. This book was written when Odoo 14 was released, and the contents are described according to the functional option of the Odoo website builder. Therefore, if you're someone who uses the latest version of Odoo, you may find the options to be slightly different than those presented in this book. You can read the references from the *Further reading* section mentioned at the end of each chapter of this book. These references describes the functional aspects of Odoo but there will be a slight change in the options described as they are based on older versions.

Download the color images

We also provide a PDF file that has color images of the screenshots/diagrams used in this book. You can download it here: `https://static.packt-cdn.com/downloads/9781801078122_ColorImages.pdf`.

Conventions used

There are a number of text conventions used throughout this book.

`Code in text`: Indicates code words in text, database table names, folder names, filenames, file extensions, pathnames, dummy URLs, user input, and Twitter handles. Here is an example: "The first way is by downloading the `.ded` file from the website and installing it before then further configuring it to be operational."

A block of code is set as follows:

```
sudo wget https://github.com/wkhtmltopdf/wkhtmltopdf/releases/
download/0.12.5/wkhtmltox_0.12.5-1.bionic_amd64.deb
sudo dpkg -i wkhtmltox_0.12.5-1.bionic_amd64.deb
sudo apt install -f
```

Bold: Indicates a new term, an important word, or words that you see onscreen. For example, words in menus or dialog boxes appear in the text like this. Here is an example: "The **Subscribe** button available for the newsletter subscription should be configured, which can be done by double-clicking on it."

> Tips or important notes
> Appear like this.

Get in touch

Feedback from our readers is always welcome.

General feedback: If you have questions about any aspect of this book, mention the book title in the subject of your message and email us at `customercare@packtpub.com`.

Errata: Although we have taken every care to ensure the accuracy of our content, mistakes do happen. If you have found a mistake in this book, we would be grateful if you would report this to us. Please visit www.packtpub.com/support/errata, selecting your book, clicking on the Errata Submission Form link, and entering the details.

Piracy: If you come across any illegal copies of our works in any form on the Internet, we would be grateful if you would provide us with the location address or website name. Please contact us at `copyright@packt.com` with a link to the material.

If you are interested in becoming an author: If there is a topic that you have expertise in and you are interested in either writing or contributing to a book, please visit `authors.packtpub.com`.

Reviews

Please leave a review. Once you have read and used this book, why not leave a review on the site that you purchased it from? Potential readers can then see and use your unbiased opinion to make purchase decisions, we at Packt can understand what you think about our products, and our authors can see your feedback on their book. Thank you!

For more information about Packt, please visit `packt.com`.

Section 1: An Overview

In this section, we'll have an overview of Odoo, the Odoo website builder tool, and learn it's used for website development.

This section consists of the following chapters:

- *Chapter 1, Introduction to Odoo and Its Website Builder*
- *Chapter 2, The Website Builder in Action*

1
Introduction to Odoo and Its Website Builder

This chapter will give you an introductory overview of the Odoo website builder along with an insight into various basic concepts regarding its operations. Additionally, we will present a brief description of Odoo and its aligned aspects of operation.

In this chapter, we will cover the following topics:

- An introduction to Odoo
- An overview of the Odoo platform
- A brief insight into the Odoo website builder tool

By the end of the chapter, you will have a clear understanding of Odoo, its benefits, and its usage within a business environment. Moreover, you will also gain basic knowledge of various operations regarding website building with the Odoo website builder.

An introduction to websites

Nowadays, the generation of business opportunities, as well as the operations of a company, mostly take place via a company website. Websites are a term that came into existence after the invention and widespread usage of internet facilities across the world. Looking back, information sharing was still a pipe dream, that is, until computer scientists, Vinton Cerf and Bob Kahn, came up with the transmission protocols that form the base of today's internet communications. The internet rose to popularity when it was opened for public use.

In the middle of 1991, World Wide Web services opened for the first time for citizens across the globe to use. We can thank the modern era of digitalization for cementing its wide usage and development. Following this period, websites came into existence and the top companies of the day began to establish their own websites to showcase their products and attract customers. Although the concept of e-commerce was established in 1979 by Michael Aldrich, a British inventor, its modern-day popularity came with the use of the internet to function with it.

In fact, it could be said that present-day e-commerce services came into existence due to the availability and accessibility of internet services across the world. With the introductory concepts of digitalization and modernization in electronics and technology, websites have retained their current design and operation. Furthermore, we could say that the advancements in telecommunication have paved the way for the widespread use of the internet and websites across the world. As the world took a turn toward the twenty-first century, internet services became an essential part of living, and they are now considered the modernized inventions of the century. Additionally, if you look back in history, the internet has paved the way for various inventions in fields that have helped human lives, and brought in advanced technologies and functioning to this world.

In a nutshell, the internet, and its associated services, can be called the invention of the century. Now, let's return to websites: With internet services being established, people discovered the widespread usage and capabilities of websites. While previously derived as a medium to share information, it was now considered as a method of communication, information exchange, and data gathering. Today, websites are used for multiple purposes such as education, business, communication, social networking, and more. Additionally, information from websites helps us with our daily tasks of decision making, information gathering, and communications.

As websites started to indulge in each person's life around the world, the usability and the need for them began to grow. Business establishments took advantage of the widespread use of the technology and began to develop their own websites. This served them well since these websites acted as a medium to showcase their products and services to the world. Additionally, as this mode of communication and company exposure was cheap and economical, it cemented the marketing strategies of these companies. Nowadays, websites and digital marketing have widespread capabilities that impact human lives. Since the majority of companies run most of their business operations through websites, designers and developers are striving to make them attractive, informative, and provide end users with a sense of satisfaction.

In this digital era that we live in, every company and institution should have a website for its operations. We can class this as an inevitable change because rather than simply being a business platform, it can be used as an informative tool and marketing platform for companies. In addition, if you are in the starting stages of your establishment, a website will be a helpful tool. Additionally, with the right design and proper descriptive content, you can drive visitors to your company website and therefore, improve your business opportunities.

Another aspect to consider is the manageability of these websites in a real-time environment. While dealing with website-based terminologies, business operations could get trickier and unmanageable without proper tools and operational methodologies being implemented. In addition, if a firm functions with a retail environment, wholesale warehouse operations that also manage website operations would need a clear-cut plan and certain operational systems in place to manage and control every operation. If you consider a website operation, you would need to design and develop it, function with it, manage the operations including in-house operations, provide attractive content and designs, track visitors, and garner business opportunities. These are the concerns to bear in mind while dealing with website operations. And the answer to these concerns is Odoo. Odoo is an **Enterprise and Resource Planning** (**ERP**) software that has been developed to run company operations. Odoo houses a website builder tool that is the ultimate solution to all of the preceding concerns. We will cover this in more detail in the final section of this chapter. However, before that, let's acquaint ourselves with what Odoo is.

An overview of Odoo and its operational aspects

With the introduction of digitalization across the world, business management solutions have started to evolve and are now operating and controlling full-fledged operations within companies. ERP terminology came into existence before this and the introduction of digitalization into the world was functioning with limited capabilities. With the modernization of technology and information exchange, ERPs evolved to be the much-needed business management software solution around the world. Odoo is one of the modernized ERPs available today and is capable of running the business operations of companies.

Let's delve into a brief insight into the history of Odoo. The founder of the platform, Fabien Pinckaers, who is currently the CEO of the company, developed TinyERP, which was later named OpenERP due to the capabilities it proposed. The company started to grow, and within a short span of time, it was able to achieve unimaginable growth compared to various other firms. Finally, in 2014, the product was renamed Odoo, which is the same name as the company.

Today, Odoo has more than 5 million users across the world and is one of the most widely used ERP solutions. Initially, it started as a software service company until transforming itself into a software publishing company. Therefore, Odoo has been with the business sector since the start, catering to every need of users across the world. Furthermore, the company started with limited employees, a handful of partners, and was developed across 100 countries. Now, Odoo has over 950 employees, 2,550+ partners, and is used over by 5 million users across the world; therefore, this ERP software has grown beyond the management and company's chart expectations.

How was it possible? Well, the only answer is the capabilities and operational smoothness that the platform brings to a company. Moreover, it offers a cost-efficient solution by using a reliable business management module to run all operations, covering distinctive aspects of the company using a modular structure. Today, Odoo is used by a vast number of industries on all sorts of operational levels. Although the platform is best suited for small and medium enterprises, the customization ability of the platform has enabled its operations to be adapted to larger establishments. The key aspect of Odoo is that the ERP is an open source platform that makes it flexible and adaptable to any change in the operational strategies of a business and its users. Moreover, this feature adds to the cost-effectiveness factor by providing more speed to operations and, therefore, improving performance. Additionally, data security aspects are more reliable and secure for operations. Furthermore, the open source platform has a network of communities, which support users with their needs.

Having understood the history and background of Odoo, let's take a look at the supporting aspects of the Odoo platform in the next sections.

The Odoo community

As mentioned earlier, Odoo is an open source ERP and has a community that supports the user in every part of operations. The Odoo community is a collection of Odoo developers, consultants, and Odoo partners. Simply put, the Odoo community can be considered the backbone of the platform, as they are responsible for bringing in regularized updates, which modify the operational aspects and add new features to the software. Additionally, Odoo releases a new version of the software every year, and it has been doing so for the past couple of years, which is made possible only by the contribution of the Odoo community. While the minor updates are being released throughout the year, the new versions are done during the Odoo Experience event.

Odoo apps

The Odoo platform allows you to use various application-specific modules of operations, which are designed and developed by Odoo community members. The applications are centered on being operational for specific functions in a company according to its needs. The users of Odoo can obtain them from the Odoo Apps Store available on the Odoo website. Some of the applications there are free of charge, while certain others come with a fee. Additionally, they are developed to smooth company operations and bring in user-friendly apps to users of Odoo.

Editions of Odoo

Odoo ERP releases a new version every year at the Odoo Experience event. The operations of Odoo mainly fall into two separate editions. Odoo ERP can be availed by customers via two editions: the first one being the Community Edition and the second one being the Enterprise Edition. Among these, the former edition is limited in terms of its features and comes free of charge. Odoo Online is one of its versions; however, its customization ability is limited and the user cannot run sophisticated business operations. Moreover, it's suitable for establishments with less than 50 users.

On the contrary, the Enterprise Edition of the Odoo platform has a license fee and subscription charges will be included when using it. It comes packed with additional features and ultimate customization ability. Furthermore, the user can run sophisticated business operations making it suitable for larger business environments. So, while Odoo ERP is suitable to function with small and medium scale enterprises, the software can be also configured to run in larger establishments. Additionally, both versions of Odoo ERP have the support of the Odoo community, which can be accessed by users for both technical and functional aspects of operations.

Odoo customization

Customization is the best feature of Odoo, as it provides the user with the ability to flex the operations of the platform as per their own requirements. Moreover, the software is developed and drafted into operations in the form of a business, based on the needs and necessities of the company's operations. In addition, the use of add-ons from the Odoo Apps Store need to be customized for each operation. In the instance of using one or more application from the Odoo Apps Store, they need to be configured to be operational with one another as well as the platform in order to satisfy the customer's application-specific needs. Additionally, Odoo customization will bring in operational features and options required for the user to run their company operations in Odoo as per their terms and conditions.

Odoo hosting

A key aspect of Odoo pricing that makes up most of the Odoo platform functioning in your company is Odoo's hosting capabilities. These can be chosen from multiple options based on the needs and necessities of the user. Odoo hosting can be done on three different types of platforms:

- Firstly, Odoo's cloud-based hosting hosts the online Odoo version, which is the Community Edition of the platform. It comes with limited features and capabilities for company operations.

- Secondly, Odoo's on-premises hosting can be done either on an in-house server or one that is available locally, according to the needs of the company. In this type of hosting, the platform can be customized along with the use of add-ons to run the company operations.

- Thirdly, the **Odoo.sh** platform can be hosted on a server provided by Odoo. It can be either a shared one or a dedicated one. The shared server is suitable for small-scale establishments, while the dedicated one is better for the larger establishments of operations.

Now that we have discussed the hosting aspects of Odoo, let's take a look at how to install the Odoo platform onto your systems.

Installing Odoo on a Linux-based system

Odoo ERP can be used in Linux after successfully implementing it to be operational. This can be done in two ways. The first way is by downloading the .ded file from the website and installing it, and then further configuring it to be operational. Another way is via command-line installation, which we will discuss next. You can efficiently perform the task in the Ubuntu system by following these steps:

1. Update the server and system to the latest version:

    ```
    sudo apt-get update
    sudo apt-get upgrade
    ```

2. Secure a server for your operations:

    ```
    sudo apt-get install openssh-server fail2ban
    ```

3. Create a user on Odoo:

    ```
    sudo adduser --system --home=/opt/odoo --group odoo
    ```

4. Next, we need to perform PostgreSQL configuration. The steps are as follows:

    a. Install postgres, as follows:

    ```
    sudo apt-get install postgres
    ```

    b. Switch to a Postgres version higher than 9.6 to create and manage the Odoo database:

    ```
    sudo su - postgres
    ```

    c. Now, create an Odoo 14.0 user:

    ```
    createuser --createdb --username postgres --no-createrole
    --no-superuser --pwprompt odoo14
    ```

    d. Assign the user as a superuser to attain more optional privileges:

    ```
    psql
    ALTER USER odoo14 WITH SUPERUSER;
    ```

e. Exit from `psql` and `postgres` using the following commands:

```
\q
```

```
exit
```

5. Install the Odoo dependencies using the following steps:

a. Install `pip3` using the following command:

```
sudo apt-get install -y python3-pip
```

b. Install the necessary packages and libraries:

```
sudo apt-get install python-dev python3-dev libxml2-
dev libxslt1-dev zlib1g-dev libsasl2-dev libldap2-dev
build-essential libssl-dev libffi-dev libmysqlclient-dev
libjpeg-dev libpq-dev libjpeg8-dev liblcms2-dev libblas-
dev libatlas-base-dev
```

c. Verify the installation along with the new dependencies:

```
sudo apt-get install -y npm
```

```
sudo ln -s /usr/bin/nodejs /usr/bin/node
```

```
sudo npm install -g less less-plugin-clean-css
```

```
sudo apt-get install -y node-less
```

6. Next, we need to perform GitHub cloning. The steps are as follows:

a. To enable GitHub cloning initially, GitHub must be installed on the server:

```
sudo apt-get install git
```

b. Change the system user to `odoo` prior to cloning, as this makes it more secure:

```
sudo su - odoo -s /bin/bash
```

c. Clone from the repository and its respective branch:

```
git clone https://www.github.com/odoo --depth 1 --branch
14.0 --single-branch.
```

d. Exit and continue with the installation:

```
exit
```

7. Install the necessary Python packages using the following steps:

    a. Install the Python packages and library using `pip3`:

    ```
    sudo pip3 install -r /opt/odoo/requirements.txt
    ```

    b. Download and install `wkhtmltopdf` for Odoo to support PDF reports:

    ```
    sudo wget https://github.com/wkhtmltopdf/wkhtmltopdf/
    releases/download/0.12.5/wkhtmltox_0.12.5-1.bionic_amd64.
    deb
    ```

    ```
    sudo dpkg -i wkhtmltox_0.12.5-1.bionic_amd64.deb
    ```

    ```
    sudo apt install -f
    ```

8. Configure Odoo and its services, as follows:

    a. A configuration file will be automatically created in the Odoo software. This will be downloaded and copied to another configuration folder, as depicted in the following code:

    ```
    sudo cp /opt/odoo/debian/odoo.conf /etc/odoo.conf
    ```

    b. Provide the necessary information regarding the file:

    ```
    sudo nano /etc/odoo.conf
    ```

    c. Update the configuration file, as follows:

    ```
    [options]
     ; This is the password that allows database operations:
     admin_passwd = admin
     db_host = False
     db_port = False
     db_user = odoo14
     db_password = False
    addons_path = /opt/odoo/odoo/addons
    logfile = /var/log/odoo/odoo.log
    ```

    d. Set up access rights to the files:

    ```
    sudo chown odoo: /etc/odoo.conf
    sudo chmod 640 /etc/odoo.conf
    ```

e. Create a directory of logs in Odoo:

```
sudo mkdir /var/log/odoo
```

f. Set Odoo's user permission inside the directory:

```
sudo chown odoo:root /var/log/odoo
```

g. Configure Odoo's services:

```
sudo nano /etc/systemd/system/odoo.service
```

h. Place the following snippet inside the file:

```
[Unit]
Description=Odoo
Documentation=http://www.odoo.com
[Service]
# Ubuntu/Debian convention:
Type=simple
User=odoo
ExecStart=/opt/odoo/odoo/odoo-bin -c /etc/odoo.conf
[Install]
WantedBy=default.target
```

i. Set up a root user for the file:

```
sudo chmod 755 /etc/systemd/system/odoo.service
sudo chown root: /etc/systemd/system/odoo.service
```

9.   Run Odoo as follows:

a. To start the Odoo platform, run the following command:

```
sudo systemctl start odoo.service
```

b. Check the status of the platform using the following command:

```
sudo systemctl status odoo.service
If active use the following URL to enter the platform.
"http://<your_domain_or_IP_address>:8069"
```

c. Inspect the log file using the following command:

```
sudo tail -f /var/log/odoo/odoo.log
```

d. If the Odoo services are to be started at the system's boot time, enter the following command:

```
sudo systemctl enable odoo.service
```

Installing Odoo on Windows

The Odoo platform can be installed on Windows-based operating systems using the following steps:

1. Download the executable file for Windows from the Odoo website, as shown in the following screenshot. The link to download it is https://www.odoo.com/page/download:

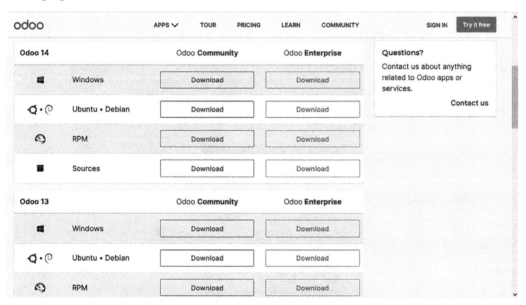

Figure 1.1 – The web page for downloading Odoo

2.   Install the downloaded file onto your system:

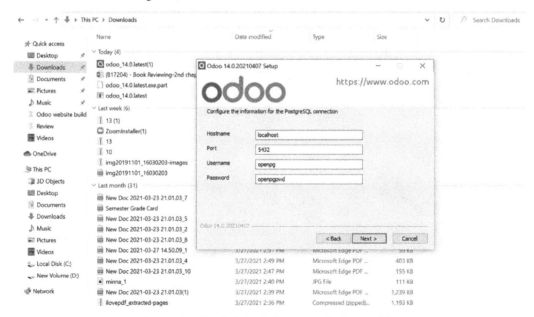

Figure 1.2 – Installation setup

3.   Confirm the PostgreSQL connection:

Figure 1.3 – Configuring information for PostgreSQL

4. Configure the destination folder:

Figure 1.4 – The destination folder configuration window

5. Start the application and configure the database for operations:

Figure 1.5 – The database configuration window

So, we are now clear on the installation aspects of Odoo for Linux- and Windows-based operating systems. Let's take a look at the key features of Odoo ERP next.

Key features of Odoo ERP

Odoo is considered one of the most advanced ERPs available today, and it is capable of running any business operations efficiently within the platform. Here are some of the key features of the Odoo platform:

- It has high-end customization abilities, allowing it to flexibly operate in any form of business operation.

- It offers single platform-based operations to control all business operations.

- It has a modular design with designated application modules of operations.

- Its integration ability allows it to operate with advanced applications and devices such as IoT and biometric devices.

- It utilizes centralized inventory management and database operations for efficient functioning along with secure data transfers.

- It has advanced operational tools such as drop shipping, cross-docking, data reconciliation, lead enrichment, and more.

- It uses a field service module to run your company's field service operations.

- It uses a fleet management tool to manage the fleet of vehicles you have.

- It provides an HR management form for the recruitment of employees to pay their salary.

- It uses a project module to run the specialized operations of your company.

- It uses a manufacturing and repair module that manages the production aspects.

- It has advanced reporting features on each of the modules of operations, providing efficient analytical and quantitative reporting on the various aspects.

- It is an efficient Odoo website builder.

- A data cleaning application is available from the latest Odoo version 14.

- It offers breadcrumb-based operations and informative details on all the sections of operations.

Note that these are some, but not all, of the features of, Odoo ERP. Lets, now move on to understand certain benefits of using Odoo.

The benefits of using Odoo

Odoo ERP brings in various advanced features, as mentioned in the previous section. Moreover, there are various additional benefits of using Odoo for users. Here are some of them:

- The extensible architecture provides the user with the ability to rewrite the code and extend the operations of the platform to new levels of company functioning.

- The cost factors of ERPs are high; however, the Odoo Community Edition can be availed free of charge, albeit with limited features, providing users with an insight into the platform's operations. Furthermore, the Enterprise Edition comes packed with features and its customization ability has a license fee that is on the lower side compared to other competitors.

- The open source platform makes the source code available to anyone, and users can access this to add their own customizations, as needed, to the code.

- It is available in two versions: the first one is the Community Edition with limited features and a lack of customization ability, while the second one is the Enterprise Edition with full-ranged features and capabilities along with full customization capabilities.

- A secure system of operations allows the user to keep their data secure, safeguard confidential and in-house data, and much more.

- The Odoo studio tool helps users to develop and modify their applications according to operational needs.

These are just some of the benefits of Odoo for users of the platform. In the next section, we will learn about the Odoo website builder tool.

An overview of the Odoo website builder

The Odoo platform offers an in-house and efficient website-building tool for its users. The designated website module of the platform allows users to run website operations from a development, design, and management perspective. The tool will help users design and create a website for a company using the simple drag and drop functionality. You might have a misconception that to develop a website, there is a need for in-depth knowledge of coding and programming. You are wrong, as the Odoo website builder helps users to create a website in no time using simple tools to function. Moreover, the user can customize their website as per their own operational needs using the various default options available. Additionally, users are provided with editing options that are operational from both the frontend and the backend of the website operations.

Every operational establishment needs a website in today's operation as the website would act as a medium to conduct business rather than the conventional ways used in the earlier days. The developers at Odoo recognized the need for a website building tool in their initial version, and they inculcated the website builder and management options. In the latest version of Odoo 14, the website builder tool is packed with operational features to deal with all the modernized aspects of website requirements. We have used the latest Odoo 14 enterprise edition throughout the book, however, some extra features wouldn't be available in the Community when compared with the Enterprise edition. Additionally, the website builder brings advanced functional options to the website. This is not only useful for users but also for visitors. The website module of Odoo also helps with the creation and management of the e-commerce platform of operations running in parallel with the website management and other company operations, such as the retail and wholesale parts of the business.

The next section will provide an insight into certain misconceptions regarding website building followed by features and operational capabilities that rectify these misconceptions. Additional benefits of using the Odoo website builder will be described along with a highlight of its salient features.

Clearing up certain misconceptions

Now, when considering the need for a website in the company, some of you might not agree to it. There might be questions or concerns, such as *why should my company need it? How can I design and develop it without help and programming knowledge? How successful will it be?* So, before jumping into further details about Odoo's website builder, let's clear up certain misconceptions regarding it.

Why do I need a website for my company?

This could be the first question that you might think about in your mind when considering a website for your company. You could also assume the need is reserved for corporate establishments and not normal-sized business establishments. Furthermore, you could say that it's not worth it. However, a website for your company, nowadays, will act as the primary source of income generation rather than an unused entity. Due to developments in telecommunications and internet facilities, people are no longer relying on others as consultants. Instead, they are more likely to search for a product or a company on the World Wide Web. Therefore, for your company to be discovered and known by people, the need for a website is both essential and inevitable.

Cost factor

Another aspect of concern for you might be the cost factor involved. However, it can be assumed from the results of various websites of other organizations that investing in a website is beneficial and provides you with better business opportunities. Additionally, while designing a website with the Odoo website builder, the operations won't cost you a penny apart from the subscription charges of Odoo.

Is it too early to implement a website?

The answer to this question is that you are always late if you decide to implement a website after the company has started its operations. In the beginning, when a company is about to start its operations, a website can serve as the best marketing tool available to inform the customers about the products and services that you are doing business with. Additionally, this could be an opportunity for the vendors to contact you to purchase the product from them.

Having cleared up the general misconceptions of creating a website, next, we will go through the capabilities of the Odoo website builder.

The features and operational capabilities of the Odoo website builder

The Odoo website builder is the ultimate tool of operations and will allow the user to design and create a website and manage the entire operations from the same platform. The website builder in Odoo is loaded with features that help the user to create attractive, user-friendly websites that can be operated from both the frontend and backend with ease. Here are some operational tools in Odoo that can help the user to create and run an efficient website:

- The intuitive system provides the user with the option for multiple text and font styles to be used while designing the website.
- The text editing tools are similar to a word processor, ensuring that text editing can be done with ease.
- It offers customization tools that can be expensed from the frontend and backend of the module.
- You can design the website using the available building blocks by simply dragging and dropping where you want them to be according to your design.
- It utilizes promotional tools such as pop-up messages, coupons, and promotional program integrations.

- You can opt for social media integration to create a wider reach with your customers.
- You can create and manage multiple websites for a company from a single platform.
- Multi-company operations can also be managed along with their websites.
- You can enable visitor monitoring and link tracking abilities for users.
- It supports Gengo and Google Analytics integration.
- You can keep in touch with your visitors using the live chat tool.
- It offers mobile previews and editing options.
- It offers frontend HTML CSS, and JS editor tools.
- You can access various professional themes by default and source additional ones from the Odoo Apps Store.

Additional benefits of the Odoo website builder

Odoo, as mentioned earlier, is a business management solutions tool with a modular structure, which has designated modules of operation. The modules of operations are interconnected and will operate in the same direction to strive for the excellence of your company. Furthermore, there are various modules of operations that are interconnected with Odoo and the website:

- E-commerce is an aspect of company operations that goes hand-in-hand with the company website, and as a matter of fact, the user, in Odoo, can define the e-commerce operations inside the company website itself, making the management operations of both run easily.
- The user can post blogs, and create forums and various other additional pages of operations in the website, which can be managed in the same production functionality. Additionally, the e-learning aspects of the company are associated with the company's website and can be provided as a separate page, which can be accessed with ease.
- The recruitment aspects of online job postings can be done on the same website and could be managed using the designated recruitment module in Odoo.

Additionally, Odoo will not limit the function and controllability of the website. Therefore, it will provide the user with capabilities to assign employees of the company to run the various aspects of the website, such as live chat, visitor monitoring, portal user allocation, and much more. Moreover, the platform exhibits various tools of operations that allow the user with functionality options to run the company website operations more efficiently on the platform.

The salient features of the Odoo website builder

Here are the salient features of the Odoo website builder that make it capable of running the entire website building and management operations for your company:

- Forget about coding and use the drag and drop features to build a website.

- You can create and brand your website yourself.

- You can monitor and turn visitors to the company website into customers using lead development.

- You can make use of systemized SEO tools.

- You can synchronize your website with Odoo's marketing tools for digital marketing.

- You can extend the website's capabilities using the add-ons available.

This section introduced you to the basic features and capabilities of the Odoo website builder. This brings us to the end of this chapter.

Summary

In this chapter, we initially learned about the emergence of websites, including a brief history of them. We also discussed the vital importance of them in today's world. We then moved on to understand Odoo ERP, discussing an overview of it, and describing the Odoo community, the Odoo Apps Store, editions of Odoo, Odoo customization, and Odoo hosting. We also learned about the installation of Odoo on Linux-based systems and Windows-based systems. Moreover, the key features of Odoo ERP and the benefits of using Odoo were described in detail. We then looked at the Odoo website builder tool, where we discussed an overview of it along with its features, its additional benefits, and the salient features when using it. Additionally, certain misconceptions regarding the Odoo website builder tool were also cleared up.

In the next chapter, you will be taken into an in-depth discussion of Odoo's website builder and learn about some of the basic tools of operation that can be used. The chapter will serve as an introduction to website building with Odoo.

Questions

1. What is the basic operational principle of Odoo?

2. When was the last version of Odoo released and what's the version number?

3. How can you design a website using the Odoo website builder?

Further reading

- *Working with Odoo* by Greg Moss, Packt Publishing
- *Learn Odoo* by Greg Moss, Packt Publishing

2
The Website Builder in Action

In the previous chapter, we learned the basics of the Odoo website builder tool. Now let's go into detail about its operations. This chapter will provide an insight into the website builder and the basic tools used to design and configure your website with ease in the application. We will learn how to create, design, and operate multiple as well as single websites for a company using the available tools, which are enough to run basic website operations in Odoo.

In this chapter, we will cover the following topics:

- Creating a new web page
- Managing pages
- Working with theme editor options

By the end of the chapter, you will be capable of creating a new web page and changing the themes of the website. Moreover, you will have in-depth knowledge of how to obtain new themes from an Odoo 14 website.

Creating a new web page

Before jumping into the web page creation and design, let's understand how to create a website on the Odoo platform using the website builder tool. The Odoo website builder module comes as standard on the platform and the user can avail of the application by installing it from the application modules available. All the website tools allocated within the website are available here and you can select them to be installed as per the requirements of your operation. You can select the **Website** module in the menu on the left side of the screen as shown in the following screenshot, and then remove all the filtering options to view the applications and associated add-ons available with the Odoo platform's website building aspects:

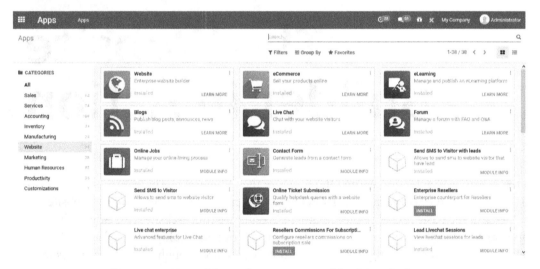

Figure 2.1 – View of the applications menu of the Odoo platform

From the menu, you can select and install the various additional modules of website operations required. Moreover, if you need to know more information about any module, you can avail of the **MODULE INFO** option available with each module.

Now as the website and associated modules have been installed, you can navigate to the website module to start the website building operation. The initial display of the website module in Odoo will depict a graphical overview of the website sales operation and the user has various provisional defaults as well as customizable options to describe the view and graphical display based on the necessity.

The following screenshot shows the dashboard of the website module in Odoo and provides you with information on the e-commerce operations. Moreover, you can navigate to the various functional menus from this window:

Figure 2.2 – E-commerce dashboard

To create a website for the company, you can select the **Settings** menu from the **Configuration** tab of the dashboard. In the **Settings** menu, you can view the **Select the Website to Configure** tab under which the new website can be created by selecting the **CREATE A NEW WEBSITE** option. Moreover, the websites created will be displayed in the **Configuration** tab and you can select the respective one to be operational if functioning with multiple websites. The **CREATE A NEW WEBSITE** option can be chosen from the **Configuration** tab as indicated in the following screenshot:

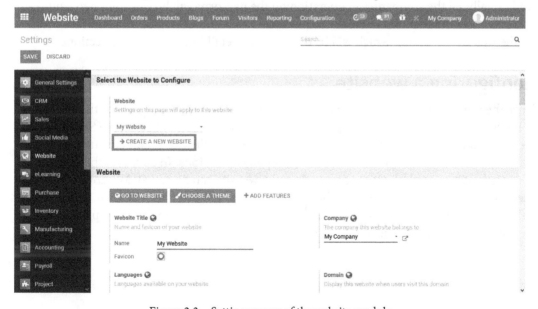

Figure 2.3 – Settings menu of the website module

On selecting the **CREATE A NEW WEBSITE** option, you will be presented with the menu depicted in *Figure 2.4*. Here, the **Website Name**, **Website Domain**, **Website Logo**, and **Company** that should be functioning details should be described. On the Odoo platform, a website can only be used for a single company, even if the platform is used to manage multiple companies:

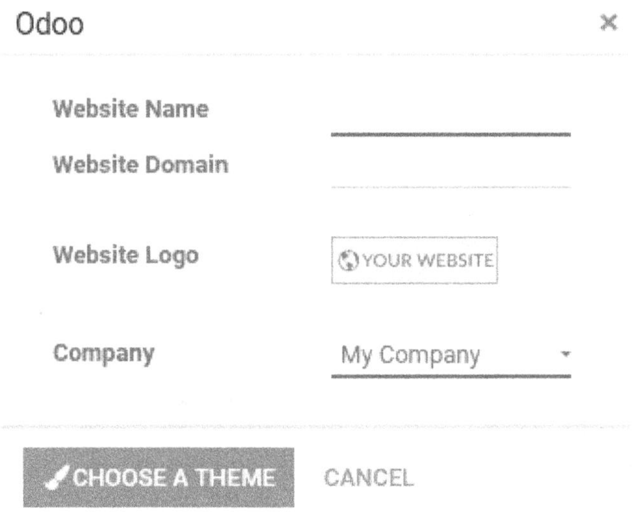

Figure 2.4 – Website creation menu

Additionally, all the companies in operation on the platform will be listed out here and the user can select the required company to function. Select the **CHOOSE A THEME** option for the website operation, which will be discussed in detail in the upcoming sections.

Configuring a website

Once the website is created, you should initially add certain go-to options for the website to function on a basic level. To assign certain basic options, you can select the **Settings** menu and under the **Website** tab, a description of the website can be added. Furthermore, the website can be made to run along with company operations. In addition, the domain of operation along with the language used on the website can be added. Moreover, the user can install new languages onto the platform if they have not been defined. The website configuration settings available in Odoo can be shaped and the description can be written as depicted in the following screenshot:

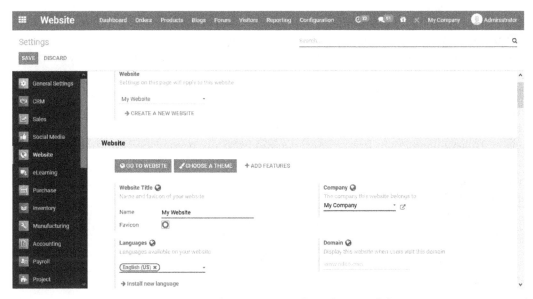

Figure 2.5 – Website settings in the website module

To add a new available language to the website, you can select the **Install new language** option and the window shown in *Figure 2.6* will be shown. Here, the respective language from the list can be selected and can be added to the available website by enabling the respective one from the list and adding it to the website:

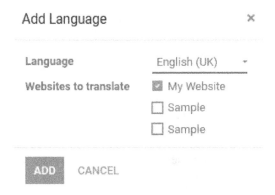

Figure 2.6 – Add Language screen

Additionally, you can describe the other options, such as using a specific user account while using the website, which only permits visitors to continue once they have created a user account on the website. Another one is the cookies bar, which can provide you with a customizable cookies bar the user can configure. In addition, social media integration can be enabled and links to distinctive social media accounts on Twitter, Facebook, GitHub, LinkedIn, YouTube, and Instagram of the company can be added. Moreover, a default social media share image can be provided, which will be displayed on the website. This can be selected from the system and from any storage location to which the platform has access. Additionally, the events PWA options can be described and a name can be provided for the events being conducted. The website's configurational tools are depicted in the following screenshot:

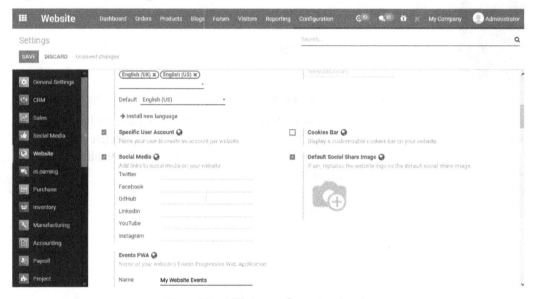

Figure 2.7 – Website configurational tools

As we are clear on creating and configuring a new web page and a website for an operation, now let's jump into the next section on how to manage the web pages created efficiently with the options available.

Managing pages

Once you have created a website, the content of the website should be defined. The content of the website is a resourceful tool to attract visitors. Moreover, the contents and pages should be defined based on the product and the services provided by the company. The user has full authority to undergo the customization aspects of the website and the constants involved with it. To view the website created, you can select the **Go to website** option available in the **Settings** menu of the website module. The website's design and specification will be as provided by you. The website's home page is depicted in the next screenshot and will be seen as based on the theme selected:

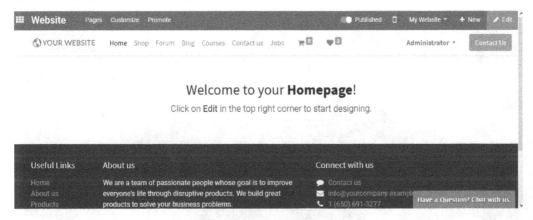

Figure 2.8 – Website home page

As we are on the website's home page, let's now move on to describing the website based on the requirements that will be described in the sections to come.

Creating a new page

You can create new pages for the website to add content. Some websites will have certain default pages; however, the Odoo platform requires you to add additional features in order to add pages, hence the available options. Here, you can create customizable pages and their content as per your needs. Moreover, numerous pages can be created to be operational. You can add new pages to the website by selecting the **New** option available at the frontend of the website and the following options window will be displayed. Here, the user can choose to add a new page to the website, blog posts, events, forums, job offers, products, courses, or appointment forms:

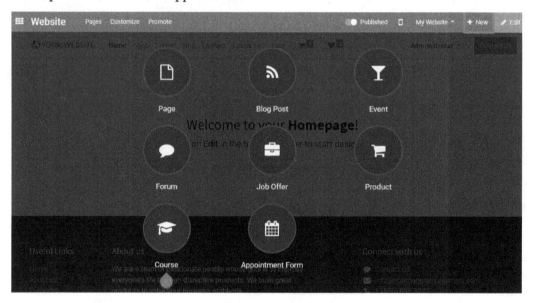

Figure 2.9 – Page addition window

However, here a web page has to be created by selecting the available **Page** option. On selecting to add a new page, the user will be presented with the following window to provide a page title and the option to enable the page in the menu or not:

New Page   ×

Page Title: Sample

                                      ⬤ Add to menu

          Continue   Cancel

Figure 2.10 – New page creation window

Once the page creation is completed, you can select the available **Continue** option. Moreover, you can also cancel the operation by opting for the available **Cancel** option. Once the web page has been defined, it will be depicted as a separate menu option on the website, as shown in the following screenshot:

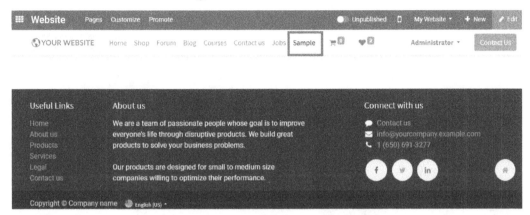

Figure 2.11 – Website home view with the new page

Moreover, you can now use the various page editing options by choosing the **Edit** option available on the respective page.

Overview of pages

The Odoo platform recognizes the need to provide the user with the ability to manage the web pages being designed and created. An overview of website pages can be viewed from the functional view under the **Configuration** tab and by selecting the page's available options. Here, all the websites described on the specified website of operation form the platform if functioning with multiple websites and multiple companies. The overview will provide you with a view name, a page URL, a website name, a description of whether the website is indexed or published, the last updated details of the user, the time and date of the update, and whether the operations of the respective web page are on track.

In addition, you can edit certain details on a web page by enabling and disabling certain options such as indexing, publishing details, and tracking information:

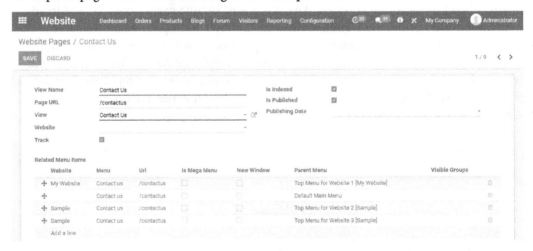

Figure 2.12 – Website page menu

Moreover, you can edit the details of the respective pages by selecting them and verifying the details available. On selecting a respective web page, you will be presented with the information menu shown in *Figure 2.13*, providing details about the view name, page URLs, view, indexing, and publishing details, along with tracking information. In addition, information on the related menu items will be provided, which will be described on the respective web page. Additionally, the related menu page information can be edited using various options such as **Mega Menu** or **New Window**. Furthermore, the respective description pages can be deleted using the delete option available for each one:

Figure 2.13 – Editing a website page

Moreover, you can add a new related menu item by selecting the **Add a line** option, where information on the website, menu, URL, parent menu, visible groups, and options such as **Mega Menu** and **New Window** can be enabled.

To create a new web page from the functional side of the platform rather than from the website, you can select the **Create** option available in the page's menu. Moreover, the user can create numerous web pages for distinctive websites and they can be interlinked in operation while functioning with multiple websites of the company or multiple companies:

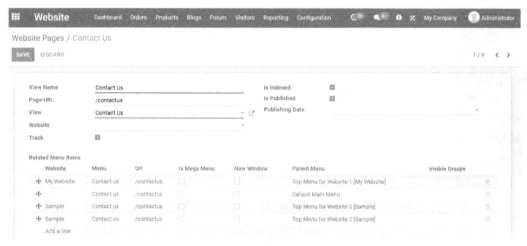

Figure 2.14 – Creating a new web page

As we have understood the manageability of web pages on your website, now let's work on the looks and design of the website. The next section will provide insight into the theme editing options in the Odoo website builder and enlighten you about how to obtain themes from the Odoo website.

Theme editor

Every website should have a definite style and functional design that make it stand out among the various websites available. Moreover, a distinctive color combination along with different styles of font and the design of the page will provide users with a soothing experience, therefore improving customer satisfaction. The Odoo platform understands this and provides you with an editing option when creating and designing a website for a company using the Odoo website builder. In addition, the theme editor functionality is easily accessible and will provide the user with certain default tools as well as customizable ones.

The theme editing option can be accessed from the **Settings** menu of the website module. And under the **Website** tab in the **Settings** menu, you can see the **CHOOSE A THEME** option:

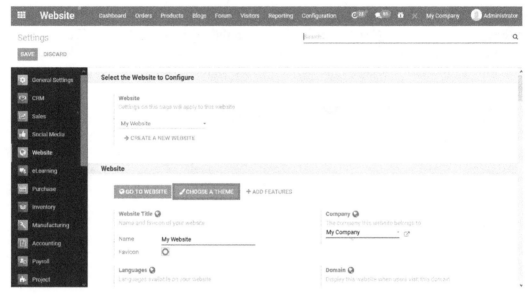

Figure 2.15 – Theme editing option in the Settings menu

Moreover, the theme and design of the website are of initial and vital importance and hence the Odoo platform will indicate to the user to choose and edit one at the time of website creation using the website builder. When you choose to create a website after providing the credential details, you can select the theme, as shown in the following screenshot:

Figure 2.16 – Theme editing option when creating a website

Another option to change the theme is from the **Editor** window of the website, as depicted in *Figure 2.17*. From the **Edit** menu, you can select the **OPTIONS** tab and select the **Theme** option from the **Website Settings** menu, upon which you will be directed to the theme menu of the Odoo platform for your website:

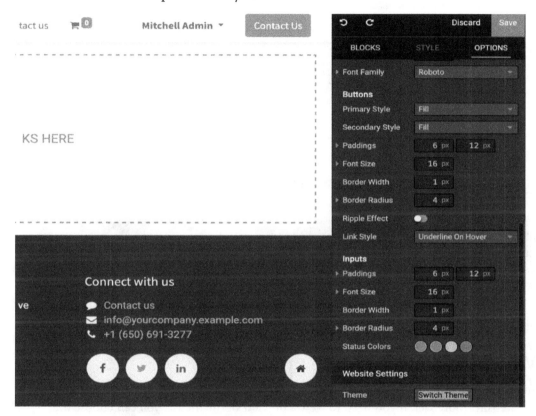

Figure 2.17 – Theme editing option while editing a website

With either of the ways described, you will be directed to the respective theme editing menu. Here, all the themes installed on the system are described and you can select any one available to operate on the website, as depicted:

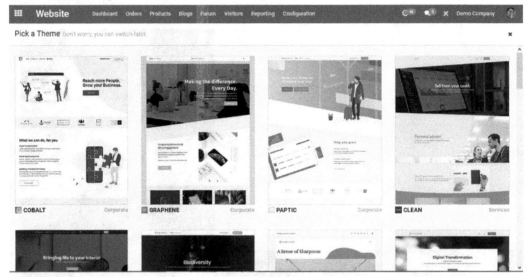

Figure 2.18 – List of themes to choose from

Moreover, the Odoo platform recognizes the need for change in a website's design and allows the user to change the theme and style as per their need.

Working with the theme preview

In addition, you can navigate to a theme and then select the theme to be operational on the website. Moreover, if you need to view a preview of the theme, you can choose the **LIVE PREVIEW** option available for the respective theme:

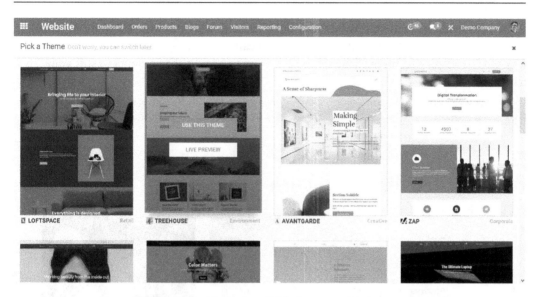

Figure 2.19 – Options available for respective themes

On selecting the respective preview of a theme, you will be directed to the preview window, where you can choose another theme to preview and make the preview be either in desktop mode or a mobile view:

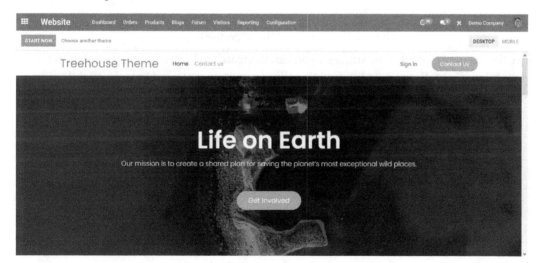

Figure 2.20 – Theme preview menu

For the preview to be loaded, it will take about 5 seconds as your entire company website should be aligned with the provided settings. You can navigate as you do on a website and check out the display features of the respective theme and edit the contents to suit your perspective:

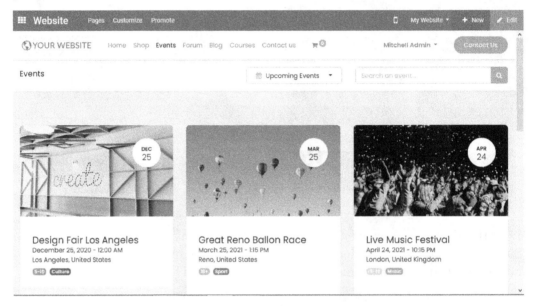

Figure 2.21 – Preview of the respective theme

Once the theme is found to be suitable, you can navigate to the preview page and select the mobile view to discover its full functionality on remote devices. Once the mobile view is enabled, you will be presented with a similar window to the one shown in the case of desktop and the user can start the demo in the preview:

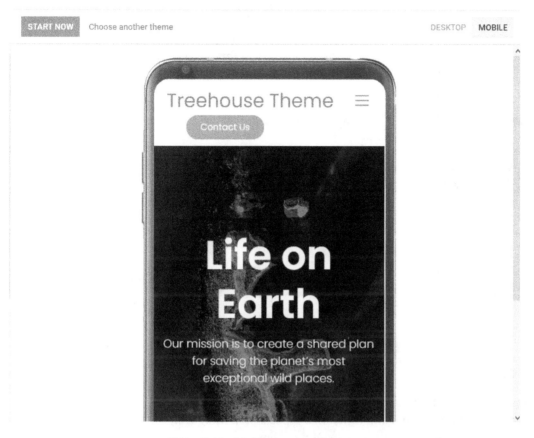

Figure 2.22 – Mobile preview of themes

You can either start the demo or choose another theme to be previewed. On selecting to start the preview in that instant, it will take around 5 seconds for your company website to be added in the respective theme of operation in the mobile view.

Let's look at another example of the operation for it to be clearer to you. From the **Settings** menu or the website creation tab, you can select an available theme option. This will direct you to a themes menu where all the themes available are shown. Here, let's select one and opt for it to be previewed before applying it to the web page:

Figure 2.23 – Choose from a list of themes

You can preview as many themes available on the platform as many times as you like to understand the design and functional aspects of it. Moreover, the desktop view and mobile view options are available in all the themes' preview windows:

Figure 2.24 – Preview menu of a respective theme

To view the desktop preview of the website for the respective theme, you can select the available desktop option and opt for the **Start Now** icon. Once the option is chosen, it will take a fraction of time for the website to be loaded up with the respective theme:

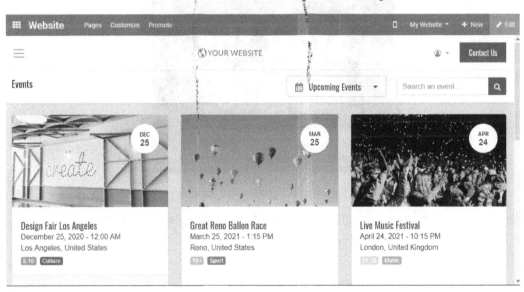

Figure 2.25 – Desktop preview of the respective theme

Navigating back to the themes preview menu, you can choose the mobile view and select the available **START NOW** option. This will provide you with a preview of the respective theme in the mobile view and how it will be depicted on remote devices:

Figure 2.26 – Mobile preview

Now, as we have understood theme selection and preview operations, let's jump into the section on how Odoo themes can be obtained.

Odoo themes

If you are a user of the Odoo platform, you can avail of various theme applications from the Odoo app store and change the filtration of Odoo themes. Here, all the themes related to being operational in Odoo will be depicted and you can avail of them by downloading the app to your platform as depicted in *Figure 2.27*. Among the themes available, there are free ones that do not have a license or subscription fee. However, there are certain paid ones for which you have to pay a license charge to the respective developers of the application:

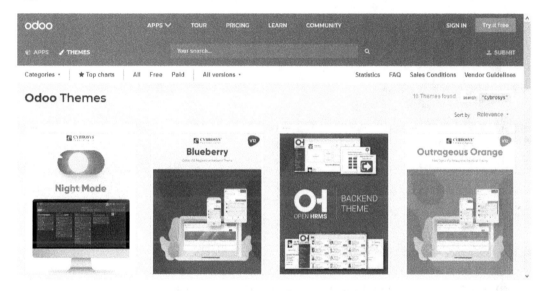

Figure 2.27 – Themes menu of the Odoo app store

You can select any of these to be applied to the platform. If the respective theme is a paid one, you should pay the license fee. The free ones can be downloaded directly.

Here is an example of the operation: you can select the respective theme and download it for the respective version of Odoo that you use. You should choose the application to be downloaded in the version of Odoo you have for it to be compatible with operations. Moreover, if you use Odoo 14, download the respective theme for the respective version by clicking on the version option shown in the following screenshot:

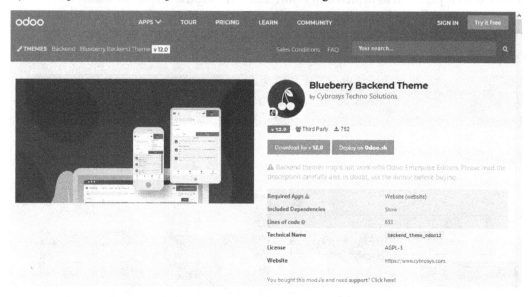

Figure 2.28 – Description window of a theme in the Odoo app store

Once the themes of the website are selected, you can run it to be operational on the platform and the website operation. Moreover, if you need to edit the theme, you can easily choose to edit it by selecting a different theme to be operational, or else you can download the new one from the Odoo apps menu and install it to be operational on the platform.

Installing Odoo themes

The operation of installing a third-party theme on your Odoo platform is quite simple. Initially, you should download or purchase the respective theme from the Odoo app store. You should ensure that you have got the right theme for the version of Odoo that you have. If you have Odoo 14, download the theme for the respective version by selecting the version icon in the respective theme window. Secondly, sharpen it to add to the menu of the Odoo platform. Finally, go to the website's **Settings** menu and select the respective theme to be operational on the website.

In addition, you can download or purchase Odoo themes from the respective Odoo partner website or their Odoo application hosting websites. Moreover, you can avail of the services of an Odoo partner for the perfect configuration of the themes and their associated content on the website.

In conclusion, the Odoo website builder's theme editor tools provide you with the ability to change the view and the design of the website in general.

Summary

This chapter provided you with an insight into the initial stages of website building using the Odoo website builder tool. Initially, we discussed how to create a new website using the Odoo website builder, then we moved onto the discretional aspects of it. Additionally, we discussed how to create new pages, and manage them effectively with the Odoo website builder. Further, we discussed the theme editor options available, and the different ways to describe themes were discussed. An insight into how to purchase new themes and add them to your Odoo platform was also given.

In the next chapter, we will dive deep into the website building aspect using blocks, where you will be provided with clear knowledge on block operations in website building and an insight into the structure of blocks and their types.

Questions

1. How do you create a new website in Odoo?
2. How do you create a web page for your website in Odoo?
3. How do you obtain additional themes for your website?

Further reading

- *Working with Odoo* by Packt Publishing
- *Learn Odoo* by Packt Publishing

Section 2: Website Builder in Depth

In this section, you'll get an in-depth understanding of the functional aspects of the Odoo website builder, with detailed descriptions of blocks and how they are used.

This section consists of the following chapters:

3
An Introduction to Blocks

So far, we have learned how to create a website and implement certain basic options such as creating web pages, managing them, adding themes, and modifying them. Now, let's move onto the various aspects of website design and appearance management using the tools that are available in the Odoo website builder.

In this chapter, we will cover the following topics:

- Gaining an insight into blocks
- Learn how to structure blocks
- Types of Structure blocks

By the end of this chapter, you will have a good understanding of the various types of Structure blocks available in the Odoo website builder, and you will know how to use them to design your web pages effectively.

Technical requirements

Since we have started to design our website, you should have already configured the website module on the Odoo platform. Alternatively, you can install the website module from the applications list available in the apps module of the platform. Note that a basic knowledge of website building using Odoo will be an added advantage.

Understanding what blocks are in Odoo website building

Designing a website will always be hectic if you don't have the ample tools to perform certain operations. The design and looks of a website matter greatly, and since your visitor is of vital importance, the website should be pleasing to the eye rather than just holding chunks of information. The ideology of website building is to keep it simple and sleek. You should avoid loading it with content and adding unnecessary illustrations as graphical content. The use of color coordination for certain operations should also be kept simple in order to please the user. Additionally, the various options available should be listed straight away and be made easily accessible. There should also be an option to choose the display in an elegant way that you require; therefore, making it suitable for you to view the content in an aesthetically pleasing way.

The Odoo website builder tool allows you to design and create your website operations. The functioning of the website builder tool is user-friendly and has a drag and drop approach. Moreover, users of the Odoo platform do not need to learn the coding or programming aspects of designing and creating a website. The website creation in Odoo is under the functional aspects of the platform by configuring the options, as discussed in *Chapter 2, The Website Builder in Action*. Now, regarding the design of the website, this can be done based on the block tools that are available. These block tools are elegant box styles, which can be dragged and dropped onto the website page in the frontend of the platform to configure the website design.

In Odoo, there are numerous block tools available that you can choose from. You can access them by choosing the edit option, which is available on each website page. Additionally, this tool is accessible from any page on the frontend of the website. There are numerous block tools in Odoo; however, in general, they have been classified into four types, as follows:

- Structure blocks
- Feature blocks
- Dynamic Content blocks
- Inner Content blocks

Each of them has various types of blocks that are available under the hood, and they will be discussed later in the book. Since we have obtained an overview of the operations of the blocks in the Odoo website builder, let's take a look at the types of blocks available in more detail. In the next section, the specifics of Structure blocks will be discussed. Additionally, the types that are included under this classification will be described, with examples, in detail.

Structure blocks

With the introduction of the Odoo website builder tool onto the Odoo platform, developers understood the need for useful tools to design a website, just as they do in the normal way, using Odoo. However, website building without Odoo requires highly technical knowledge and an in-depth understanding of programming and coding. The Odoo platform found this unnecessary and had to provide a better solution for website building. Therefore, the Odoo platform introduced the blocks system, where you can simply drag and drop different operational blocks to design your website. In addition, these block tools are available for all the web pages on a website.

The Structure blocks available in the website builder of Odoo include a collection of basic tools that are necessary when website building. There are basic blocks of operations that can be dragged and dropped into the website to bring in a basic structure to the website and allow you to prepare your content accordingly. Additionally, the Structure blocks in the Odoo website builder are easily accessible from the blocks window. Furthermore, it is important to note that all the website building operations can be done from the frontend of the platform, and this will be described in further detail in this chapter and the upcoming ones.

To access the block tools, you can make use of the edit option available on the frontend of the website. To do this, you can enter the website to be designed from the go-to option available in the backend of the platform. Once you enter the website, you will be able to view the **Edit** option available in the upper-right corner of the screen, as shown in the following screenshot:

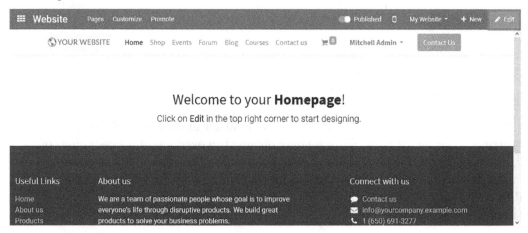

Figure 3.1 – The home page of your website

All of the block operations and website editing tools can be availed using the **Edit** option. Note that the **Edit** option can be found on all the website pages, and it can be used in a similar manner, providing you with similar editing options and tools to design the website in an eye-catching way. After selecting the **Edit** option, you will be presented with a menu on the right-hand side of the screen containing all the options available for you to design the website using the Odoo website builder. The menu will be categorized into sub-menus, including **BLOCKS, STYLE,** and **OPTIONS**. The **Structure** block and its allocated types are depicted under the **BLOCKS** menu. There is a search bar for the **BLOCKS** window, as there are numerous options available, so you can easily filter out your desired one. The following screenshot depicts the **Edit** window in the website builder, and the **Structure** block section is also highlighted:

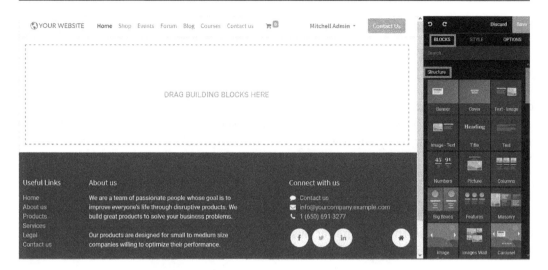

Figure 3.2 – Website editing menu and structure blocks

There are around 18 styles of structure block tools available that you can choose from to carry out your website design, as follows:

- The Banner block
- The Cover block
- The Text - Image block
- The Image - Text block
- The Title block
- The Text block
- The Numbers block
- The Pictures block
- The Columns block
- The Big Boxes block
- The Features block
- The Masonry block
- The Image Gallery block
- The Image Wall block
- The Carousel block

- The Media List block
- The Showcase block
- The Parallax block

You can add numerous blocks to a web page to create the design of the website as per its needs. Now that the concept of Structure blocks and how to obtain them in the Odoo website builder is clear, let's jump to the types of Structure blocks and learn how to design a website with them.

Types of blocks

We are now going to learn about the various types of blocks that are available in the Odoo website builder. We will start with the Banner block.

Banner block

In the list of the types of Structure blocks used in website design with Odoo, the Banner block comes first and is one of the best design tools available in the Odoo website builder. The Banner block provides a distinctive banner for your website, which you can place on any available web page. To access the **Banner** block option, select the **Edit** icon from the frontend of the website, and navigate to the list of **Structure** blocks in the **BLOCKS** menu; the banner icon is the first option that is depicted. Take a look at the following screenshot to see how to access the **Banner** block:

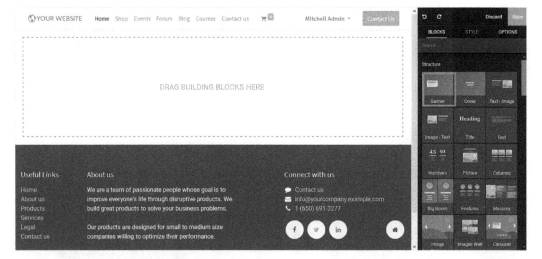

Figure 3.3 – The Banner block

Once the Banner block has been selected, you can drag and drop the respective block to the web page where the design will suit this particular operation according to your requirements. Note that as the Banner block is implemented on the web page, a default style of the respective block also appears on it. The contents of this block can be edited from the **STYLE** menu, which you can access from the right-hand editing menu of the web page. While the default content of the text, the background image, and any sort of available media can be edited, the style of the block will remain the same:

Figure 3.4 – Preview of the Banner block in action

To change the content of the Banner block, you can easily navigate to the options that are available and change them as needed. The following screenshot shows an example of a web page after changing the current text:

Figure 3.5 – Editing the content in the Banner block

Additionally, there is a sorting option that can be configured. You can use this option after providing a name and an operation as per your requirements. In most instances of website design, each different option will take visitors to a particular page, which will be configured to deal with a particular operation. After selecting to edit the respective option, you will be shown a configuration pop-up window. Here, the link to the web page or a URL can be included. The option can be configured as a **Link**, a **Primary** web page, or a **Secondary** web page based on the operation. Additionally, the **Size** and **Style** of the icon can also be configured from the drop-down list of available options. You can preview the icon on the right-hand side of the window, enabling you to view the edited style and other design aspects of the icon. Once the configuration of the respective option has been decided, you can save it to be functional on the website; otherwise, you can discard the operation. The following screenshot shows the **Link to** window where the smart option is configured:

Figure 3.6 – The window to add a link to the smart icon

After editing the operation, the respective icon is now available; you can view the changes directly on the web page and further modify them if required:

Figure 3.7 – Editing the smart icon

Once the basic editing of the Banner block has been completed, you can undergo major design changes using the editing menu that is available. Since the block we have chosen is a banner, we can now configure the various style options that are available.

The style editing tools within blocks

First, you can configure the background of your web page, which you can set as an image, a video, a filter, or a shape. This mainly depends on your preference. For all background media operations, you can also configure the position, the parallax of the view, the content width, and the height of the media from the various default options, as shown in the following screenshot.

You can do this by selecting the respective menu. Remember that this configuration can be easily changed or removed:

Figure 3.8 – The STYLE editing menu

As you can see in the preceding screenshot, there is an option to enable or disable the Scroll Down Button, which can be useful for a viewer if there is more content available on the web page. Next, let's go on to explore the various options available to configure the block style.

Adding a background image

If you want to change the background image, you can select the **Replace media** option, and then the **Select a Media** window will appear. Here, you can either upload a new image to be the background from your external storage, or you can add a URL where the image can be uploaded from:

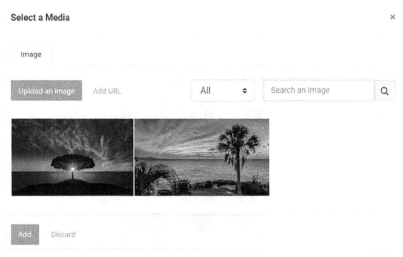

Figure 3.9 – The Select a Media menu

In addition to this, all of the saved images within the platform will be displayed, and you have the option to search for an image and filter it out using the default options. Once you have selected an image, and you have verified that it should be added to the web page, you can click on the **Add** button and it will be added to the web page. Now, we can turn to the operations on the respective image. These include options to change the position of the image, the filter used, the width of the image, the quality of the image, and the parallax of the image inside the viewer. In addition, both **Content Width** and the **Height** of the content can be adjusted, which will be shown in the final image. The scroll-down button can also be enabled. You can configure additional options for this, too, such as the **Color** and the **Spacing** of the scroller:

Figure 3.10 – The resulting web page after editing

Adding a background video

The Odoo platform recognizes the need for a website to have a video playing in the background in order to engage and entertain visitors. Therefore, inside the Banner block, you can add a video file that will be played as the background. To do this, select the video icon that is available in the block **STYLE** editing menu. You will be shown a new **Select a Media** window. Here, you can add the **Video code** URL from an external video source such as YouTube, Vimeo, Dailymotion, or Youku:

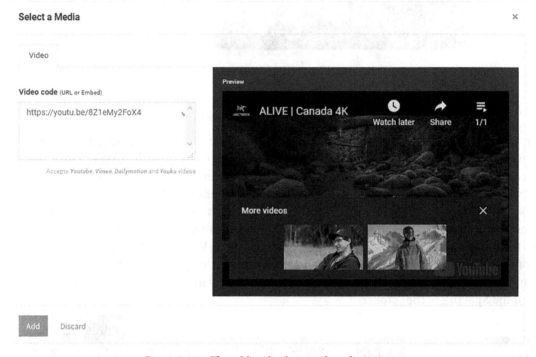

Figure 3.11 – The adding background media menu

This feature allows you to choose video content from social media for promotional purposes. Once the video URL is provided, you can click on **Add** to add the file to the website. Back on the web page, the added video will be playing in the background. In addition, the selected video can be replaced by clicking on the **Replace media** option inside the **Video** tab:

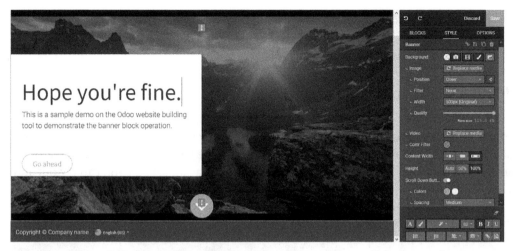

Figure 3.12 – The resulting web page after adding a video to the background

As you can see in the preceding screenshot, the **Content Width** and the **Height** of the video file can also be configured. Furthermore, the scrolling option can be enabled and the color and the spacing of the scroll icon can be configured, too.

Adding a filter to the video

You can also add a filter to the video playing in the background. This **Color Filter** option will provide you with even more customization options in which to design the website as per its style of operation. You can avail yourself of the filtering option by clicking on the paint brush icon next to the video file icon. After you have selected it, the **Color Filter** option will be displayed in the editing menu:

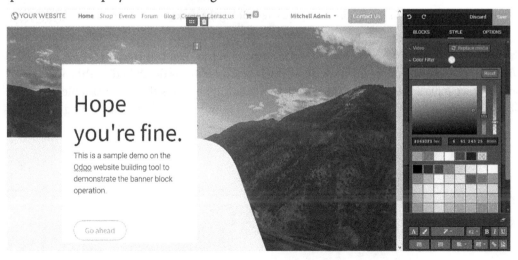

Figure 3.13 – The resulting web page after editing the background video filter

Here, you can set the color filter of the video file to be displayed in the background. There are numerous color coordination options to choose from, and you can opt for the most suitable one according to your requirements.

Adding a shape to the video display

The background video or image should have a defined display style, and the Odoo platform enables you to configure the shape of the display from the various default options that are available. You can enable the **Shape** option for the background video and choose from a list of default shapes. There are numerous choices available that you can try out and change if you do not find anything suitable:

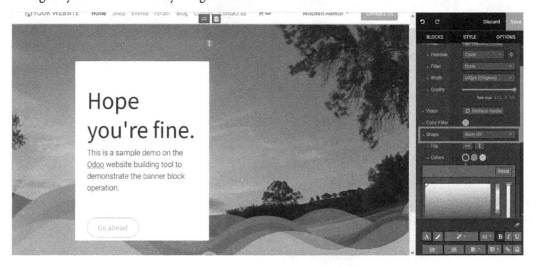

Figure 3.14 – The resulting web page after editing the background video shape

The **STYLE** editing tools can be found under each and every block operation of Odoo. The options will be similar and have unique additions, too. You can refer to the *Style editing tools within blocks* section for all of the block operations described in this book. Now that we have a clear idea of the various style editing options that are available, let's move on to the editing options that will help you to configure the block accordingly.

Editing options

There are numerous editing options that are available for your content. You can view them in the **OPTIONS** menu after selecting the **Edit** icon; you will find that you can configure various options. These options offer illustrative design tools that you can use to configure the web page; all of the available options are described in further detail in the following sections.

Title

You can edit the title of your content on a web page, including the header content and the body and the footer. You can click on the **Title** editing option, and configure the appearance of the **Background**, **Text**, **Headings**, **Links**, the **Primary Buttons** colors, and the **Secondary Buttons** colors to be displayed on the website:

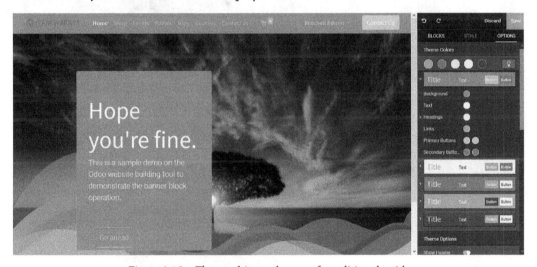

Figure 3.15 – The resulting web page after editing the title

Now that the title has been edited on the web page, let's discuss how to edit the layout of the page.

Page layout

The layout of the web page can be configured to become operational on your website. You can do this by choosing from the various default options that are available. You can select whether you want the page layout to be full, boxed, framed, or postcard style. The following screenshot shows the page layout in the postcard style:

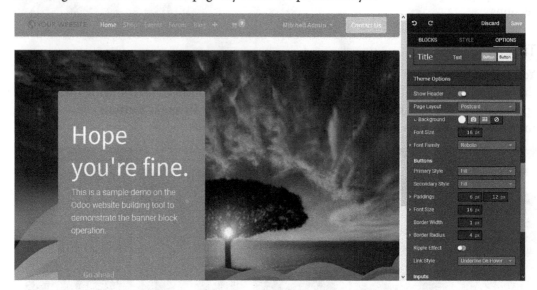

Figure 3.16 – The resulting web page after editing the Page Layout

Now that the web page layout has been edited, let's move on to the content editing options. In the next section, first, we will describe the font editing tools.

Font

The font of the content of the website is a vital aspect and will impact the presentation of the web page. The Odoo platform recognizes the need to use various styles of the font in each operational web page and has provided a number of provisional options that you can configure for this purpose. The Font Size and Font Family can be chosen for a particular web page from the list of default ones available on the platform. This will allow you to design the website according to a distinct style and appearance. The following screenshot highlights the different Font size and style options in the editing menu:

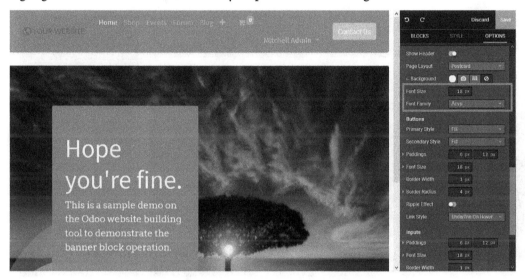

Figure 3.17 – The resulting web page after editing the font

There are various font options to choose from, and they can be implemented with ease. Next, let's take a look at how to edit the appearance of the web page.

Appearance

The appearance of the website is vital, and the Odoo platform provides you with various tools to configure the design and appearance of your web pages. There are operational tools for the configuration of the **Paddings** size, **Font Size**, **Border Width**, **Border Radius**, and **Link Style**. In addition, you can enable or disable the ripple effect of the operation. Font editing can also be done for the input operation of the content, such as **Paddings**, **Font Size**, **Border Width**, and **Border Radius**, and **Status Colors** can be configured, as follows:

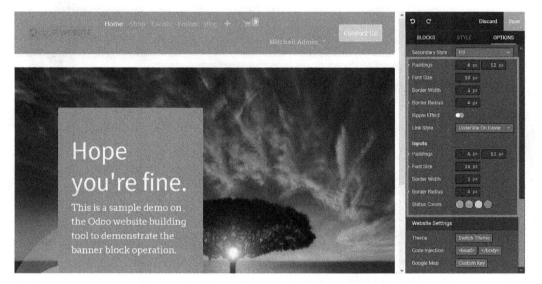

Figure 3.18 – The resulting web page after editing the appearance of the page

The appearance of the web page can be edited as per your requirement using the default tools, which will make the web page more attractive. Now we are clear on all of the editing options available in the Odoo website builder, let's move on to the editing of the website settings.

Website settings

The Odoo platform offers you editing options to configure the website both from the frontend and the backend of the platform. The **Website Settings** menu is accessible from the **OPTIONS** tab and will provide you with various operational tools to configure your website settings.

Themes

The backend editing of the themes was discussed in *Chapter 2, The Website Builder in Action,* in the *Theme editor* section, which you can refer to. You can configure the editing of the themes at the frontend by selecting the edit option on a web page and then scrolling down down the **OPTIONS** menu to reach **Website Settings**. To change the theme of the website, you can select the **Switch Theme** option, as highlighted in the following screenshot:

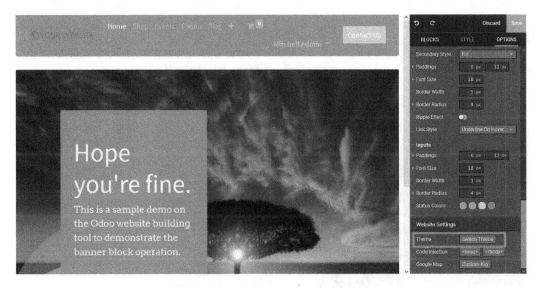

Figure 3.19 – Editing the theme in the Website Settings menu

However, the website theme cannot be changed while undergoing website editing in the Odoo website builder, so you will be shown an alert message that asks you whether you want to continue with theme editing. Therefore, the changes made in the website editing operations will be saved; otherwise, you can cancel the operation. On choosing to continue, you will be taken to a themes menu, as shown in the *Theme editor* section of the previous chapter. Here, all the themes related to Odoo are displayed, and you can opt to change to a new one. In addition, there are various preview options that are available, and you can operate them in the same way as described in the *Theme editor* section of the previous chapter. The following screenshot shows the theme selection window in the Odoo website builder:

Figure 3.20 – Selecting a theme

Once you have selected a theme to apply, it will take a while to configure the existing website to the new theme format. Once it's done, you can see that the configuration is complete, based on the respective themes along with the editing that was performed in the website design:

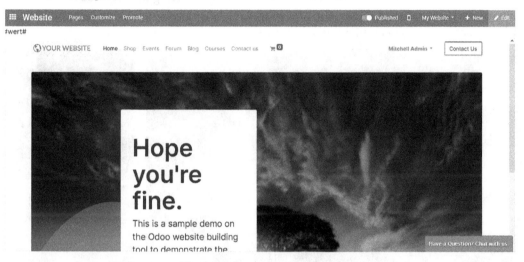

Figure 3.21 – The resulting web page after editing the theme

The theme editing options are very beneficial, and you can download or obtain new themes for your platform from the Odoo app store, as described in the *Themes editor* section of the previous chapter.

Code injection

The **Code Injection** option will allow you to insert a line of code at the beginning of the content. In the Odoo website builder tool, you can insert a code of operations in the header and the body. The injection can be customized by selecting the required option from the **Website Settings** menu. The following screenshot shows the **Code Injection** option in action; it has been highlighted in the **OPTIONS** window:

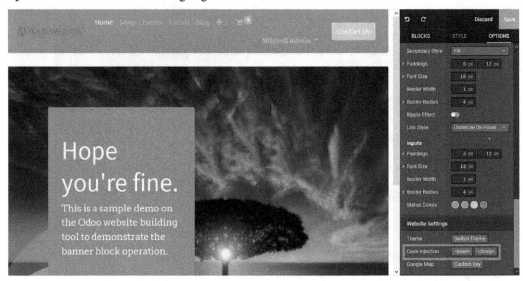

Figure 3.22 – The Code Injection options in the Website Settings menu

When selecting this option, you will be presented with a **Custom head code** menu, where you can provide customized code that will be operational within the website's content:

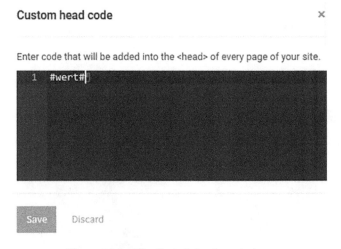

Figure 3.23 – The Code Injection window

The Google Map API

Odoo provides the option to integrate Google Maps with the website in the website builder application. This allows you to add a Google map to provide the geolocation of a company or a facility under the establishment. The Google Map API key can be accessed by clicking on the **Google Map** configuration option from the **OPTIONS** tab:

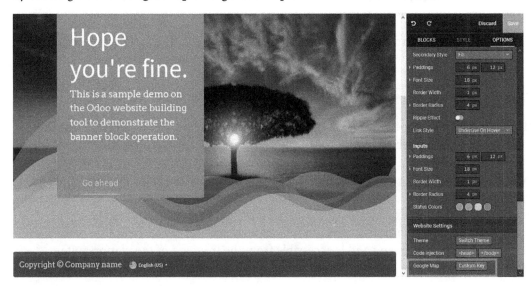

Figure 3.24 – The Google Map configuration option

By selecting the **Google Map** option, you will be presented with the **Google Map API Key** menu. Here, you will have options to create a Google project and obtain a key upon subscription; alternatively, you can enable the billing option on your Google project and copy the available key:

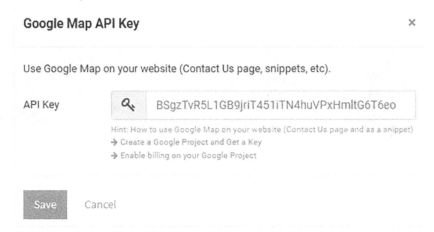

Figure 3.25 – The Google Map API Key window

The **STYLE** and **OPTIONS** menus of all the blocks available in Odoo will be similar, and you can easily configure them during each block selection.

Since the options are similar in each of the blocks, they will not be described in the following block description. Instead, you can refer to the *Banner block* section. Furthermore, if there are specific different options available for a block, it will be described in the relevant block section in upcoming chapters.

Now that we have a clear idea of the Banner block and its operations, let's move on to the next section, where we will learn about other block operations in Odoo.

Cover block

The Cover block is a type of design tool that you can use during website building to add a catchy cover to the web page. The Cover block option will provide provisional space in which to assign a catchy title to the website and can be placed at the top of the web page to be displayed upon the visitor's entry into the website. The Cover block functionality will provide you with the contact details for the company. In addition, the contact details can be provided on an external or separate web page and the link to the respective web page can be displayed here. The **Cover** block can be found under the **Structure** block description of the **Edit** menu and is accessible from the frontend of the platform. It is located right next to the **Banner** block icon:

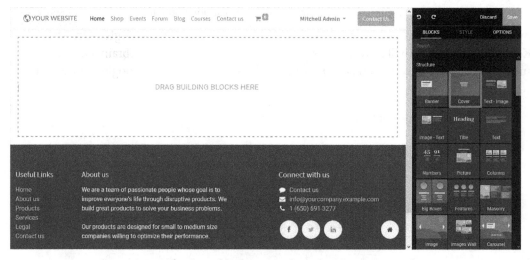

Figure 3.26 – The Cover block icon in the web page editing window

You can easily drag and drop the block to the web page to add it to the design. Note that its appearance will be based on the theme selected, and the color coordination will also be displayed accordingly:

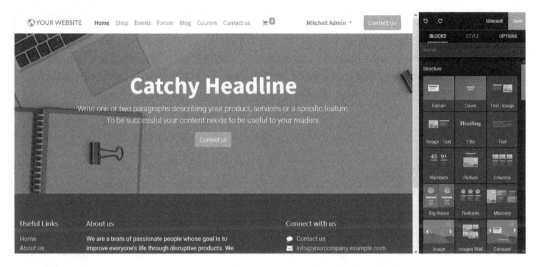

Figure 3.27 – A web page preview of the Cover block

You can provide a title for the Cover block and allocate a description to it. This can be customized by selecting the content on the web page and editing and modifying it according to your needs. Here, the content is modified to be a suitably catchy heading along with additional content related to it:

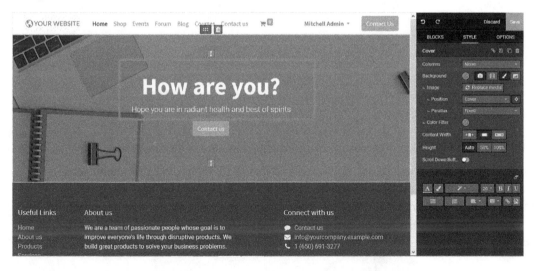

Figure 3.28 – The resulting web page after editing the Cover block

Additionally, there are smart button options configured in each block, which can be used to direct the visitor to a separate page containing specific details. Usually, this type of page is used to provide the contact details of the company, or it can be set up as an inquiry form to allow customers to share their concerns or feedback. Alternatively, it can be any specific page that the company wishes the customer to be directed to. By double-clicking on the icon, you will be presented with an editing menu, where you can configure the link. Here, the URL or email of the operation can be configured. Additionally, the page anchor, the type of link, its size, and its style can also be configured. You can either enable or disable the option to open the link in a new window, as shown in the following screenshot:

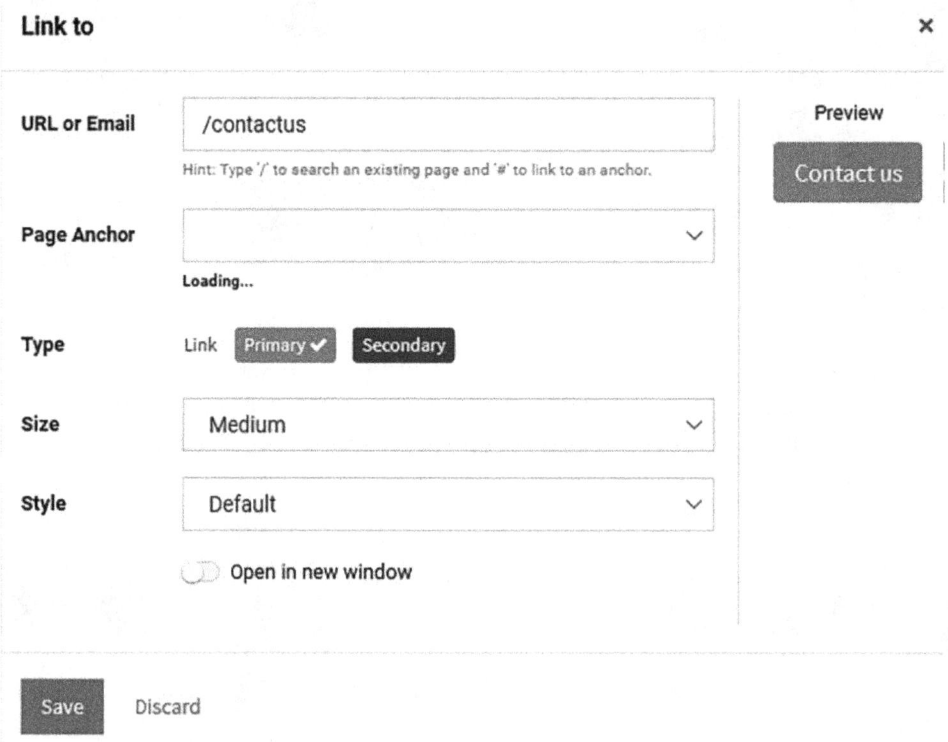

Link to   ✕

			Preview
URL or Email	/contactus		
	Hint: Type '/' to search an existing page and '#' to link to an anchor.		Contact us
Page Anchor		⌄	
	Loading...		
Type	Link Primary ✔ Secondary		
Size	Medium	⌄	
Style	Default	⌄	
	⬤ Open in new window		

Save   Discard

Figure 3.29 – The smart icon configuration window of the Cover block

The Cover block is a basic block operation and its **OPTION** and **STYLE** menu settings can be configured in a similar way to the Banner block. If you have any concerns regarding it, you can refer to the *Banner block* section. Now that the various operations regarding the Cover block are clear, let's jump to the Text - Image block in the next section.

Text - Image block

The Text - Image block design is useful for inserting text content along with an image into the web page. This is one of the most used designs and can be widely seen on websites. In the Odoo website builder, the block is the third in the list of **Structure** blocks. Because the Text – Image block brings a kind of structure to the website, it is classified under the Structure block category. The Text - Image block can be chosen from the list of structure blocks after selecting the edit option on a web page during website building.

Once the block icon is selected, it can be dragged into the web page and dropped in the required location. The design of the Text - Image block contains two sections that are configured to include a text description and image placement. Both of these can be edited and removed when needed. In addition, you can also add sections to the respective block as per your requirements. The following screenshot depicts the default web page of the Text - Image block:

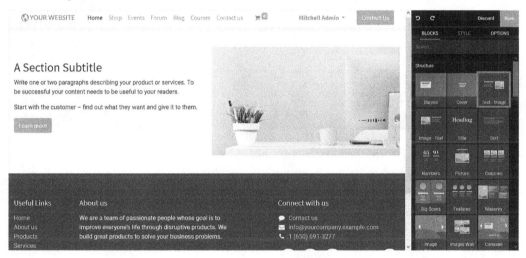

Figure 3.30 – Web page preview of the Text - Image block

The contents variable can be provided by the user and customized accordingly. The text content that you use will be as per your requirements and according to the company's products and services. You can provide an image of the product, the company's operations, or any default image available in the system. All the options for editing and the various style operations are available, as described in the *Banner block* section. The following screenshot shows an example of the resultant web page after both editing and configuration are done on the Text - Image block:

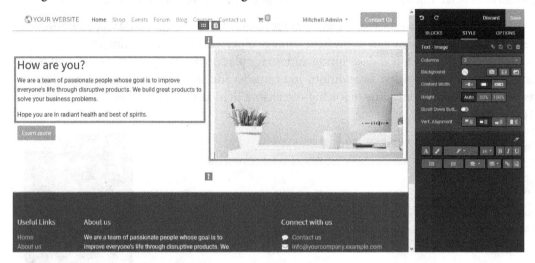

Figure 3.31 – The resulting web page after editing the Text - Image block

Additionally, there is another option in which you can direct the visitor to a different web page or a separate page on the website. You can configure this option by double-clicking on the **Link to** button, and a new window will appear:

Link to	✕

URL or Email	#	Preview
	Hint: Type '/' to search an existing page and '#' to link to an anchor.	Learn more
Type	Link Primary ✔ Secondary	
Size	Medium ⌄	
Style	Default ⌄	
	⬭ Open in new window	
Save Discard		

Figure 3.32 – The smart icon configuration window under the Text - Image block

Here, an email for direct mailing operations or a URL for an external link can be added. Additionally, the icon Type, Size and the Style can be specified. All the configuration options and styles regarding this block operation are similar to that of the *Banner block* section. You can refer to it to gain a clear understanding of the same.

Now that we have a clearer understanding of the Text - Image block of the website builder, let's jump to the next available Structure block: the Image - Text block.

Image - Text block

This Structure block type is available in the Odoo website builder tool and provides you with a design functionality, using blocks, to add an image and some content to your web page. The Image - Text operational block is similar to the Text - Image one; however, the appearance of the block is reversed. The Image - Text block can be chosen from the **Structure** block list from the **Edit** menu that is available in the right-hand corner of the web page. After selecting the Image - Text block, you can drag and drop the block into the web page in the precise location that you want it to be. The design of the block includes an image section and a text content section that can each be edited and configured:

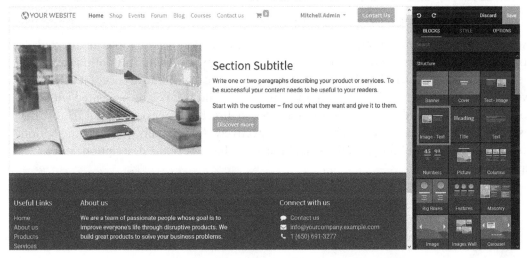

Figure 3.33 – Web page preview of the Image - Text block

Note that this image can be replaced. You are provided with various provisional options in which to edit the image. In addition, the textbox can be configured and updated with a customized one as per your requirements:

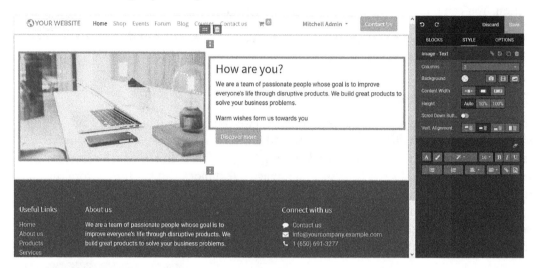

Figure 3.34 – The resulting web page after editing the Image - Text block

Additionally, there is an option that you can configure to allow the visitor to be directed to a respective window or an external page of the company's website. This option can be configured to be operational in the platform, and you can edit the details of it as you wish. To do so, double-click on the **Link to** smart icon, and you will be presented with the following pop-up window. Here, the desired URL to direct the user or an email ID that the user can send their queries to can be added, respectively. Additionally, you have the option to configure the type of the icon, the size of the icon, the style of the description, and its appearance:

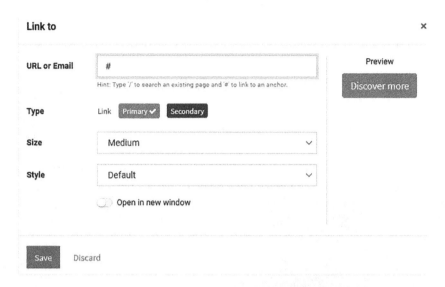

Figure 3.35 – The Smart icon configuration window underneath the Image - Text block

We are now clear on the design of the Image - Text block and we know about the various editing and style options that are also available to customize the block. These options are similar to all the other block operations in this list, as detailed in the *Banner block* section.

In the next section, we will gain an understanding of the Title block and its operations.

Title block

The Title block is one of the basic website building operational structural blocks and is widely used when building the website of a company. The Title block will provide you with a structure to design and build a title on the website. The Title block option can be obtained from the **Structure** block section, which is available in the editing menu of the website builder.

Once the block has been selected, it can be added to the web page according to a precise location. You will see a **Title** block structure that provides a distinctive title for your operations:

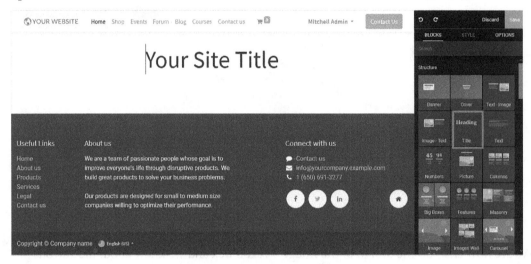

Figure 3.36 – Web page preview of the Title block

In addition, the title available can be edited and customized to suit the website operations of the company. Furthermore, there are various editing tools available, such as **Columns**, **Background**, **Content Width**, and **Height** editing options:

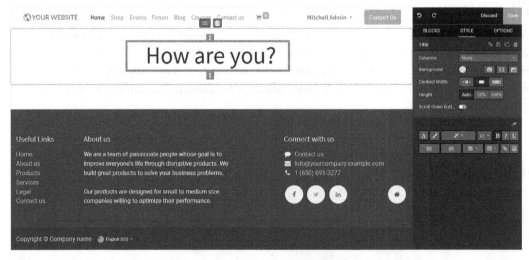

Figure 3.37 – The resulting web page after editing the Title block

Here, the **OPTIONS** and **STYLE** editing options are similar to the Banner block, which you can refer to in the earlier section. Now, we have a clear understanding of the Title block and its configurations. So, let's jump to the next block operation in the website builder: the Text block.

Text block

The Text block is a structure-based block tool for website design where you can provide text content of any length on the website. This tool is beneficial for web pages that have more descriptive content. The Text block can be chosen from the list of **Structure** blocks available in the editing menu of the website builder, as shown in the following screenshot:

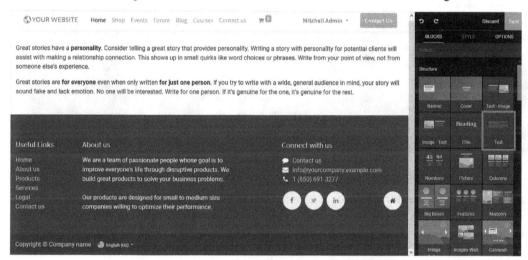

Figure 3.38 – A web page preview of the Text block

Once the Text block has been selected, you can drag and drop it to the desired location on the web page. The Text block is accessible from all of the web page editing windows of the Odoo website builder. Initially, there will be default content available that can be modified to suit your requirements. The content can be of any length or number of words, and they can be further edited using the editing tool whenever this is needed:

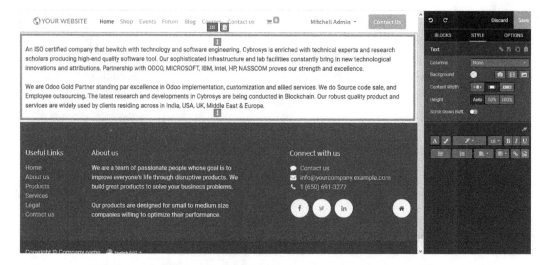

Figure 3.39 – The resulting web page after editing the Text block

Additionally, there are numerous operational **OPTIONS** and **STYLE** options to choose from, just as there are for the other building blocks of the website builder. You can refer to the *Banner block* section to gain a better understanding of the same, as all of the block tools in Odoo have the same editing and configuration options.

With this understanding of the Text block operations in the Odoo website builder, let's move onto the operations of the Numbers block in the next section.

Numbers block

There are certain instances within a website where numbers need to be highlighted to showcase the capabilities of a company. The Odoo website builder provides you with a provisional numbers block, which will allow you to put forward numerical content as opposed to text content on a web page. The **Numbers** block can be accessed from the **Structure** block menu in the website builder editing menu:

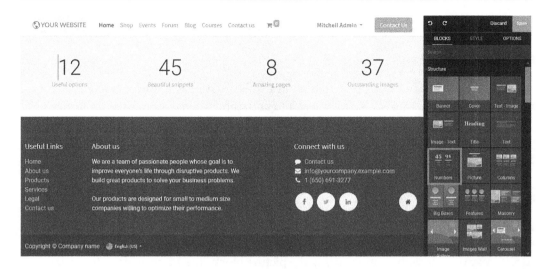

Figure 3.40 – A web page preview of the Numbers block

Once the **Numbers** block has been selected, you can drag and drop the block into the web page at its required location. In this instance, default numbers are being described along with a small amount of text-based content, which can be edited to suit you. The numbers and the associated text can be changed within the block, and you are provided with various options to change the styles and the settings of the description in the block. These options can be easily accessed from the editing menu, as follows:

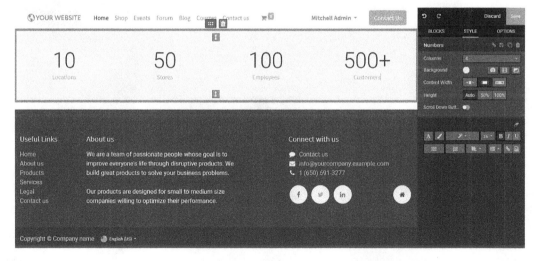

Figure 3.41 – The resulting web page after editing the Number block

The numbers and the allocated text can be changed in the block and you are provided with various options to change the style and the settings on the description in the block. These options can be easily accessed from the editing menu.

Moreover, the style and options editing tools available are similar to the ones available in the respective blocks available. If you need to know about the operation of these tools read the Banner block being described in the book. As the operations of the number block and its functioning of design are clear, we will now jump onto the picture block in the upcoming section.

Picture block

The pictures on a website will make it more attractive and eye-catching for visitors. The Odoo platform allows you to define various pictures on the web page, and you can edit the pictures available with captions using the picture block. The **Picture** block is accessible from the **Structure** block list of the website editing menu.

Once the Picture block has been selected, it can be dragged and dropped to the desired website location. The Picture block will contain options in the web page to include a heading, content in the heading, a picture, and a caption for the picture provided. All of these aspects can be customized, and you can edit the details whenever needed:

Figure 3.42 – A web page preview of the Picture block

A customizable caption is included, the picture can be edited, and a caption for the picture can be added to create a distinctive-looking web page. Additionally, there are various tools available for you to configure the block operations, as shown in the following screenshot:

Figure 3.43 – The resulting web page after editing the Picture block

Furthermore, the available **STYLE** and **OPTIONS** editing tools to deal with the **Picture** block are similar to the ones available in all the block descriptions, and you can learn more about them by referring to the *Banner block* section.

Now that the picture block operations are clear, let's move onto the Columns block and its various operational tools in the Odoo website builder.

Columns block

The **Columns** block in the Odoo website builder will provide you with a descriptive column for your web page design. The **Columns** block can be accessed from the **Structure** blocks of the editing menu. Once the **Columns** block has been selected, you can drag and drop the column to the web page, which initially provides you with three columns of image and text content to be edited. You can change the image and the content:

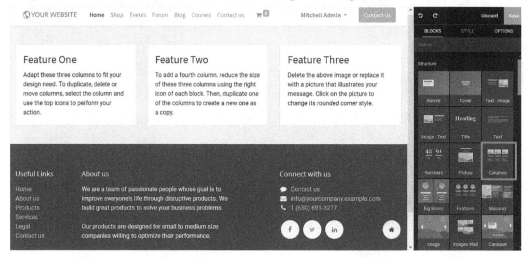

Figure 3.44 – A web page preview of the Columns block

An image from an external server or your own device can be added to the column and the same is acceptable for the content too. Moreover, you can edit the details whenever needed and remove them from the web page with ease:

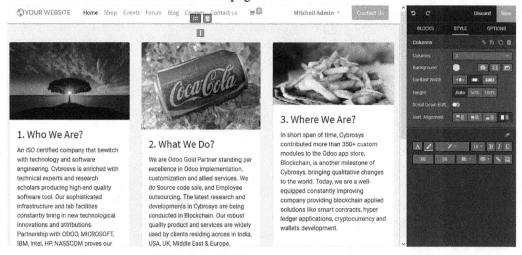

Figure 3.45 – The resulting web page after editing the Columns block

Additionally, there are various editing tools available in which to configure the styles and options regarding the contents and image in the column. Furthermore, these tools are similar to the ones available in the Banner block and all of the other block operations that are available in the Odoo website builder. If you need additional knowledge of these options, you can refer to the *Banner block* section.

We are now clear on the operations of the Columns block. In the next section, we will discuss the Big Boxes block.

Big Boxes block

Sometimes, on a website, you need to provide your content in an illustrative way as well as an attractive way. Therefore, the Odoo platform offers the Big Boxes block tool, which can be availed from the website editing menu of the **Structure** block. The **Big Boxes** block will provide you with ample space for operations on a website in which to add customized content. Once the **Big Boxes** block has been selected, you can make it operational by dragging and dropping it onto the web page. A default content and design block will be added, and you can select to edit the ones you need:

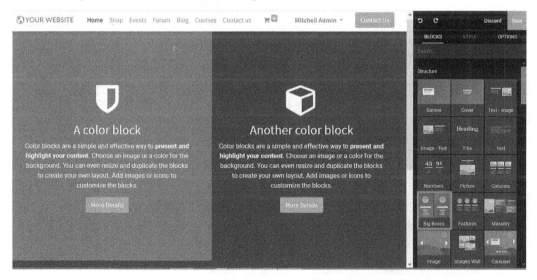

Figure 3.46 – A web page preview of the Big Boxes block

Additionally, there are options for you to select the image that you use within these boxes. A media file can be selected, or you can add one from your system. The media file can be an image, a document, a pictogram, or a video:

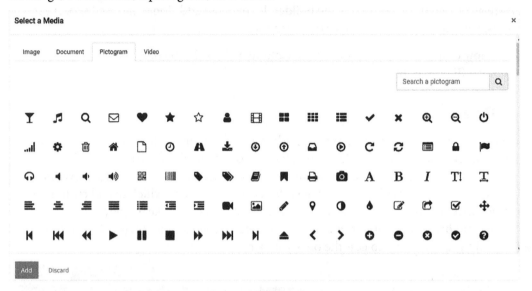

Figure 3.47 – Media selection and uploading menu

In addition, there are configurable options in the form of smart icons available in the respective boxes of the block. These can be used to direct visitors to an external web page or to an email creation window that will be sent to the company. The link type can be configured along with the size and style of the option you have chosen:

Figure 3.48 – The smart icon configuration window under the Big Boxes block

You can also edit the captions of the images along with the descriptions in the boxes. There are various optional tools available for you to configure the contents of the Big Boxes block, in the same way as the other blocks within the Odoo website builder. The options that are available are similar to the ones described in the *Banner block* section, and you can refer to that section for further details:

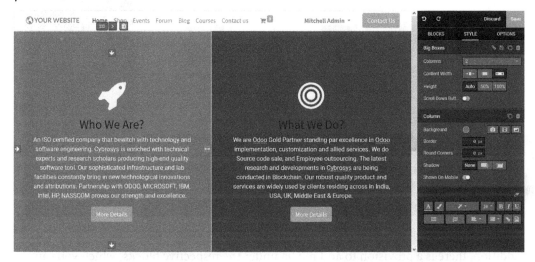

Figure 3.49 – The resulting web page after editing the Big Boxes block

Now that we are clear on the operations of the Big Boxes block, let's move onto the Features block.

Features block

The Features block found in the list of structure blocks will provide you with provisional space on the web page to describe the features of a product or service. The Features block can be obtained from the editing menu underneath the **Structure** block list.

After selecting the **Features** block, you can drag and drop it onto the web page in the location in which it will be used. Default descriptive content is provided, which can be modified and replaced:

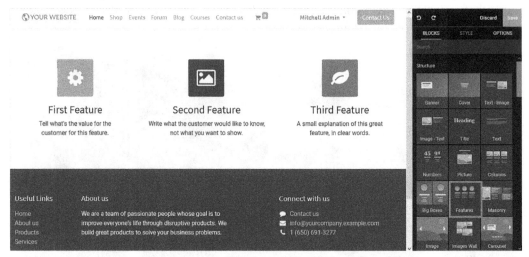

Figure 3.50 – A web page preview of the Features block

In addition, there is a provision to add media under the respective blocks, which you can also modify. To change or modify the media in the block, you can double-click on the media icon and upload an image, document, pictogram, or video for the operation:

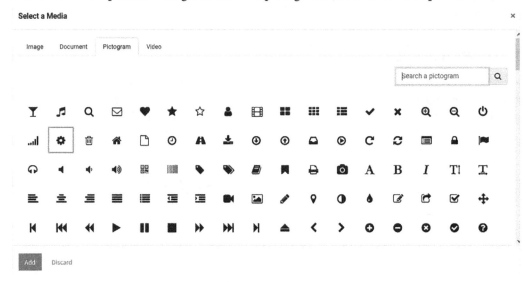

Figure 3.51 – Media selection and uploading menu

Note that the content available for each of these sections can also be edited and replaced along with the heading of each added feature. There are various options and style choices that you can configure to run the block operation on the web page as per your design requirements:

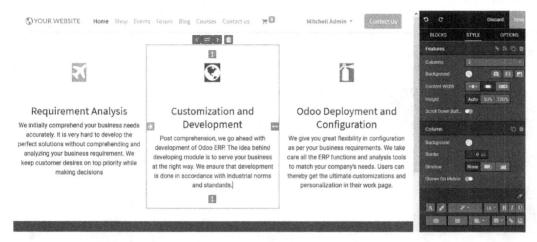

Figure 3.52 – The resulting web page after editing the Features block

The style and configuring options available are similar to the ones available in all of the block operations within the Odoo website builder. If you require a better understanding of the theme, please refer to the *Banner block* section of the chapter.

Now that we are clear on the Features block, let's go on to the next block style available under the Structure block list: the Masonry block.

Masonry block

The Masonry block of the Odoo website builder can be used to provide additional tiles for context illustration on a website page. The Masonry block option can be accessed from the list of Structure block types in the website builder tool. After selecting the Masonry block icon, you can drag and drop it onto the web page. Initially, you will be presented with default content and images inside the block, and you can adjust the configurations and edit the details in it as per your website's requirements:

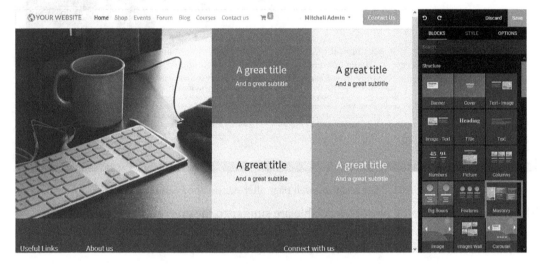

Figure 3.53 – A web page preview of the Masonry block

The image provided can also be edited and changed as per your need. You can import it from an external device or an operating system that is available to the Odoo platform. In addition, there are provisional settings and options available regarding the block, as you will find in all the blocks of the Odoo website builder:

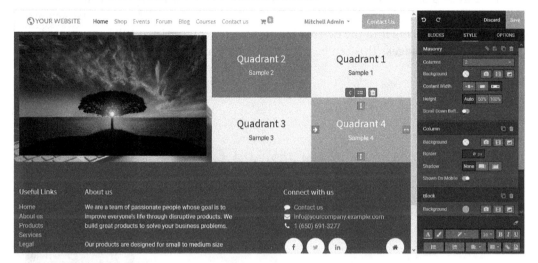

Figure 3.54 – The web page after editing the Masonry block

The options and styling tools available with respect to the Masonry block are similar to the ones available in the Banner block; you can learn more about these operations in the *Banner block* section. Now that we are clear on the operations of the Masonry block, let's move onto the next section, where we will describe another type of Structure block: the Image Gallery block.

Image Gallery block

Having images on a website is a useful tool for companies to showcase their product and services. The Odoo website builder provides you with an Image Gallery block, where you can upload images on various aspects. They can also act as a marketing tool as well as an insight into the company's products and services. After selecting and dragging the Image Gallery block and placing it in the desired location on a web page, you will be presented with a field to **Add Images**. This can be used to upload your images to the website:

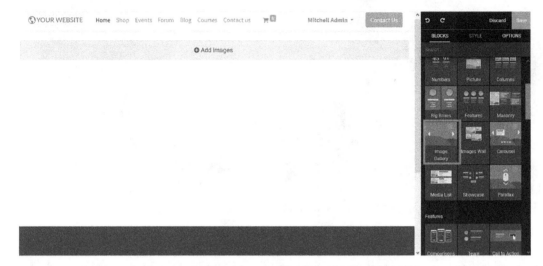

Figure 3.55 – A web page preview of the Image Gallery block and the Add Images option

After adding your image, you will be presented with a **Select a Media** window where you can upload images and add a URL for a respective image. Note that multiple images can be added to the web page via the menu:

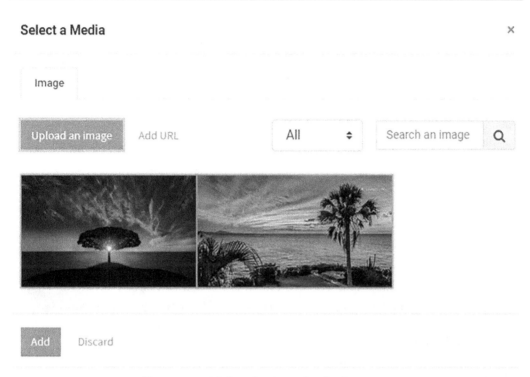

Figure 3.56 – Media selection and uploading menu

Once the images are added, you can select the **Add** option and the web page will deposit the images in an orderly fashion in which there is a scrolling option available:

Figure 3.57 – The resulting web page after editing the Image Gallery block

In addition, there are various optional and provisional styles available in the Image Gallery block, which you can edit to suit your requirements. The editing options and styling options are similar to the ones available in the Banner block, and you can refer to that section to gain a better understanding of the operations of these various options.

Now that we are clear on how to use the Image Gallery block tool within the Odoo website builder, in the next section, let's take a closer look at the Image Wall block.

Image Wall block

On a website, companies should be able to showcase their products and services, and images are a good method by which to put forward these products and services to visitors. From an image of the product or service, visitors will be able to understand their capabilities and operations in a visual manner. The Odoo website builder offers an Image Wall block tool that will provide you with a structure block on the web page to showcase your images.

The Image Wall block can be found under the category of **Structure** blocks in Odoo, and you can make it operational via the editing menu of the web page. Once the Image Wall block has been selected, it can be dragged and dropped to the web page locations where you need it. The Image Wall block tool is available across all web pages in a website designed using the Odoo website builder:

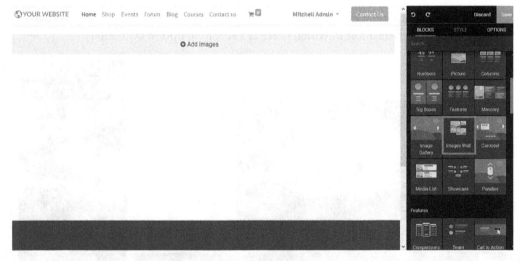

Figure 3.58 – A web page preview of the Image Wall block and the Add Images option

By selecting the **Add Images** option, the media selection menu will appear where you can add or upload new images to the platform. Furthermore, the images available can be chosen to be displayed in the block. The platform also allows you to add image URLs to obtain the image directly from the external web pages. The following screenshot depicts the media uploading window where images can be added to the platform:

Figure 3.59 – Media selection and uploading menu

Once the images to be added are selected, you can click on the **Add** option and the selected images will be displayed in your desired location in the form of tiles:

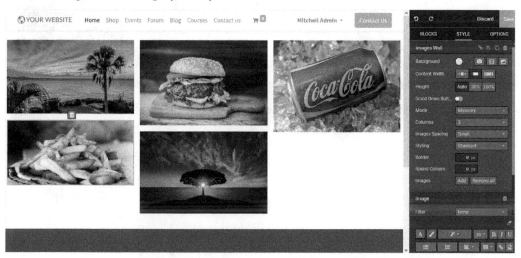

Figure 3.60 – The web page after editing the Image Wall block

There are editing options available to configure the images and their background, as per your needs, which is similar to various operations described in the *Banner block* section.

Since we are clear about the operations of the Image Wall block, let's now move onto the Carousel block in the next section.

Carousel block

On the website of certain companies, you will have seen web pages where information can be viewed without changing the view. This methodology is called a carousel view. The Odoo website builder has a Carousel structure block for you to design your website accordingly. It can be obtained from the editing menu of the web page under the list of **Structure** blocks. Once the block has been chosen, you can drag and drop it to make it operational on the web page. Initially, you will be shown three carousels to scroll between. All of them can be edited and modified to suit your needs. Also, the carousel slides can be added in any number and removed as needed:

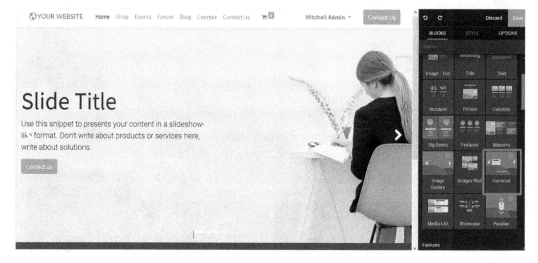

Figure 3.61 – A web page preview of the Carousel block

Furthermore, both the image and the text can be modified, and there are various editing tools available that will allow you to change them to suit your web page. There are options to change the **Background**, **Image**, **Position**, **Filter**, **Width**, and **Quality**. Additionally, there are configurable styles and options to choose from, which are similar to the other blocks in the website builder, as described in the *Banner block* section. You can refer to this section when considering the various styles and options that are available:

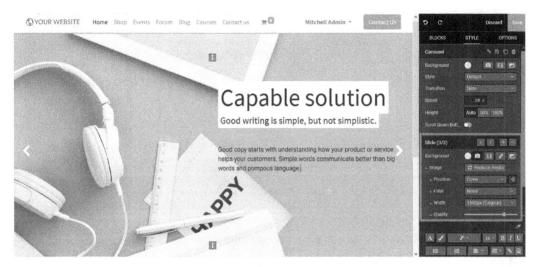

Figure 3.62 – The resulting web page after editing the Carousel block

As we are clear on the carousel block of the Odoo website builder tool, let's now move on to the media list block in the next section.

Media List block

If you require a list of descriptive content to showcase various media communications or your company's operations, the Media List block in the Odoo website builder will be very beneficial. This block is accessible from the editing menu of the web page, underneath the list of **Structure** blocks, and it will provide you with a block design to illustrate the various media content, as per your requirements.

Once the Media List block has been selected, you can drag and drop the block to your desired location on the web page. Initially, you will be provided with a default media list block. This includes a description of informative content along with an image and a smart button to drive visitors to a different web page or location on the website:

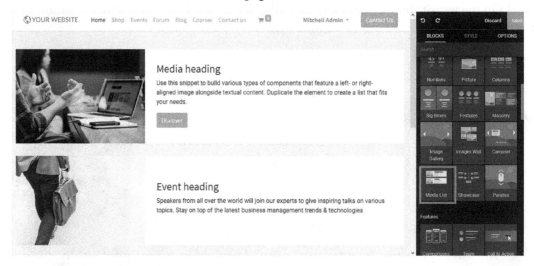

Figure 3.63 – A web page preview of the Media List block

The image and the content can be edited, and you can add customized ones. There are also default tools for in which to modify the content, which will provide you with options to change the **Background**, **Border**, and other image editing-related tools:

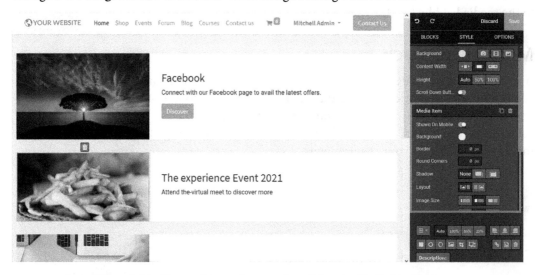

Figure 3.64 – The resulting web page after editing the Media List block

Furthermore, there are options to change the style of the block and its appearance. This is similar to the ones described in all the block operations of the Odoo website builder. You can refer to the *Banner block* section for more information.

Showcase block

The Odoo website builder provides another functional block tool known as the Showcase block. This enables you to describe the specific operational methodologies of the products or services of your company in an illustrative pattern. The Showcase block comes under the **Structure** blocks category and can be accessed from the editing menu. After selecting the Showcase block, you can drag and drop it onto the web page at a specific location. It will provide you with a default description and design that can be modified:

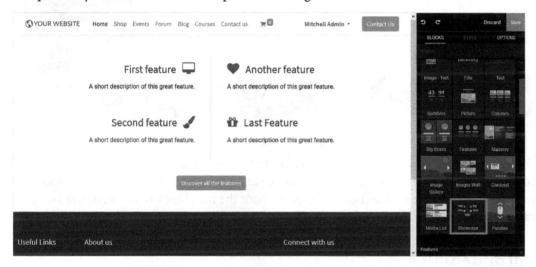

Figure 3.65 – A web page preview of the Showcase block

Both the content and the image icons can be edited and modified as needed. Additionally, there are block-specific editing options available, which you can obtain from the **STYLE** menu. Furthermore, similar to all of the blocks in the Odoo website builder, there are various style and editing options that will allow you to configure the block on the web page accordingly:

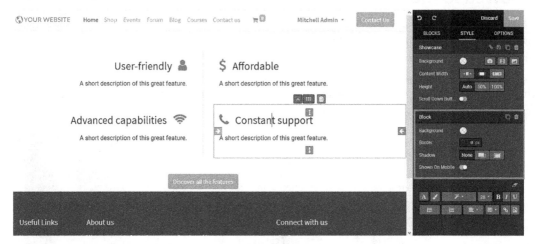

Figure 3.66 – The resulting web page after editing the Showcase block

You can read the *Banner block* section to gain a better understating of the styling and editing options that are available. Since we are clear on the Showcase block of the Odoo website builder, in the next section, we'll learn about the final block in the Structure blocks category: the Parallax block.

Parallax block

The Parallax block is a website building block that is available in the Odoo website builder. It is usually used at the end of the web page to provide an illustration image of the company's operations. The Parallax block is found in the list of **Structure** blocks and can be accessed from the editing menu. After selecting the Parallax block, you can drag and drop it onto the desired location on the web page. In addition, a default image according to the theme selected will appear, which can be edited, changed, or modified:

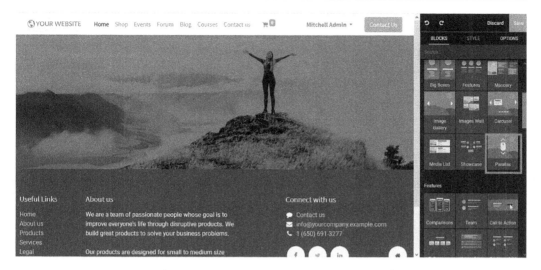

Figure 3.67– A web page preview of the Parallax block

Note that the image can be changed, and there are various editing and styling options available, similar to all the other blocks. Please refer to the *Banner block* section for more details. The following screenshot shows a preview of the resulting web page after the Parallax block has been edited:

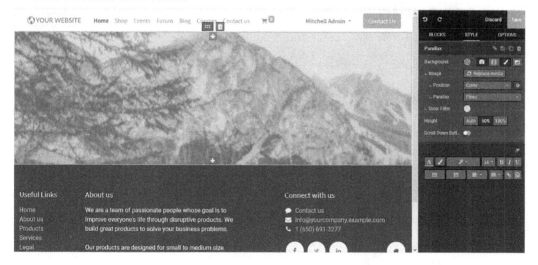

Figure 3.68 – The resulting web page after editing the Parallax block

In conclusion, the various structure blocks available in the Odoo website builder will provide you with ample options to configure your web page according to your design and company operations. Moreover, they will bring structure to your web pages and can be altered and modified as needed.

Summary

In this chapter, we learned about block operations in Odoo and how they can be used to design websites. Additionally, a detailed insight was provided for you to understand the structure block operations and the various types of Structure blocks such as the Parallax block, the Media List block, the Cover block, the Banner block, and more. We also detailed the settings options and the style configuration tools. Now, you should have a greater understanding of this topic and be able to design a web page using the structure blocks and tools associated with them.

In the next chapter, we will focus on features blogs, their operations, and the types that can be used in Odoo's website building applications.

Questions

1. What are the different categories of blocks available in the Odoo website builder?

2. How many types of Structure blocks are there? Can you name a few of them?

3. Name some of the style and editing options available in all of the block operations.

Further reading

- *Working with Odoo* by Greg Moss, Packt Publishing
- *Learn Odoo* by Greg Moss, Packt Publishing

4
Design Using Features Blocks

You may now have an understanding of the block operations in Odoo, especially the structure blocks and their various types of functioning. Although structure blocks are used to provide a structure to the website design, they are also the most widely used compared to other block categories. However, to have a website structure and content that highlights certain aspects, a structure block would not be effective. Therefore, Odoo has allocated a separate category of features blocks, which will be explained in the chapter with examples.

In this chapter, we will be covering the following content:

- Exploring features blocks
- Understanding types of features blocks

By the end of the chapter, you will be able to design your web page in Odoo in an illustrative and descriptive way using the features blocks in the Odoo website builder.

Technical requirements

As we have already covered the first type of block tools available in the Odoo website builder – the structure blocks – you will be familiar with the website building operations with them. However, for a beginner, a basic knowledge of Odoo and website building with it would be beneficial in operations. In addition, a system with an installed Odoo platform is necessary to run the operations.

Exploring features blocks

Like the structure blocks in the Odoo website builder, the features block is a category of blocks consisting of block designs that are useful in highlighting the website content. Moreover, this would provide visitors with an illustrative description of the content available. Additionally, you can easily configure them and make modifications to suit your design and style of the website. Unlike the structure block, which provides a design structure to the web pages, the features block can be used to highlight specific features of the company, its products, and services.

In addition, there are various types of features blocks (to be exact, 13 features block tools are available in the latest version of Odoo). This will provide you with ample design tools to describe the contents in a descriptive as well as illustrative way. Although the design and structure of the features block vary from that of the other category of blocks available, the design and editing options available remain the same – similar to the structure blocks as mentioned in the *Banner block* section under the structure block type in *Chapter 3, Introduction to Blocks – Structure Blocks*.

The features blocks are mostly used to describe the specific aspects of a product and services. Moreover, it will help you to describe and highlight the specific features and capabilities of the product. In addition, there are blocks under features blocks that will help you to describe the timeline of a product or a service in steps that will make it visible to the reader in a more simplified manner rather than describing the contents.

As we are clear on the usage of features blocks, let's understand the different types of the features block available in the Odoo website builder.

Understanding types of features blocks

In general, the features blocks are said to be the ones that describe the contents required in a website in a more illustrative and understandable way rather than sticking with the descriptive contents. Moreover, in total, there are around 13 types of features blocks available in the Odoo website builder, and this is more than enough to design your website.

Types of features blocks

Here is a list of all the features blocks available in the Odoo website builder:

- Comparison block
- Team block
- Call to action block
- References block
- Accordion block
- Features grid block
- Table of content block
- Pricelist block
- Items block
- Tabs block
- Timeline block
- Steps block
- Quotes block

The following sections will describe each type in detail and will be illustrated with screenshots of operations.

Comparison block

The **comparisons** block tool is the initial type of features block available in the Odoo website builder. The product comparison, as well as the comparison of your company services with that of your competitor, will be beneficial for your website visitors. The comparison block can be obtained from the website editing window available to you upon selecting the edit options available on every web page of your website, as shown in the following screenshot:

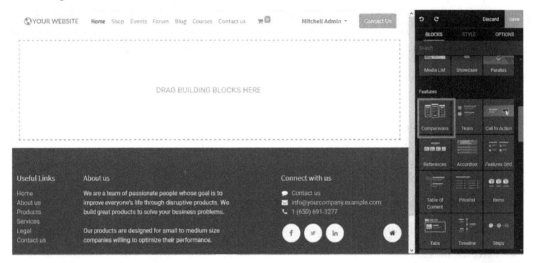

Figure 4.1 – Comparison block icon in the web page editing window

Once the block is selected, it can be dragged and dropped to the desired location in the web page. Moreover, by default, the block will only provide you with three descriptive tables to compare. These can be edited and modified to suit your operations. Furthermore, the following screenshot provides a preview of the comparison block in action in a web page:

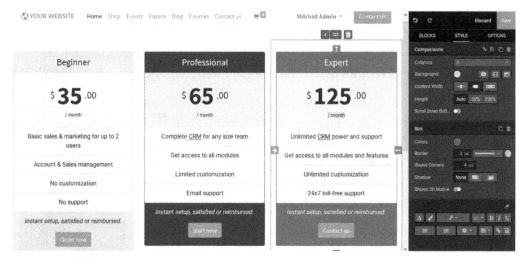

Figure 4.2 – Web page preview of the comparison block

In addition, there are configurable smart buttons that will direct your visitors to an external web page or create an email straight to your sales or customer support team. In the smart button configuration window, the URL or the email ID can be configured along with the type of the button as well as the size and style of the button. Additionally, the preview of the button being configured is available over the right side of the window, as shown in the following screenshot:

Link to

		Preview
URL or Email	/contactus	Start now
	Hint: Type '/' to search an existing page and '#' to link to an anchor.	
Page Anchor		
Type	Link [Primary ✔] [Secondary]	
Size	Medium	
Style	Default	
	◯ Open in new window	

[Save] Discard

Figure 4.3 – Smart icon configuration window under comparison block

Moreover, the comparison table and its content can be edited or removed and the allocation can be configured. There are also options to configure the style and contents of the block as similar to the ones available in all block operations of the Odoo website builder. The following screenshot depicts the resultant web page after comparison block editing:

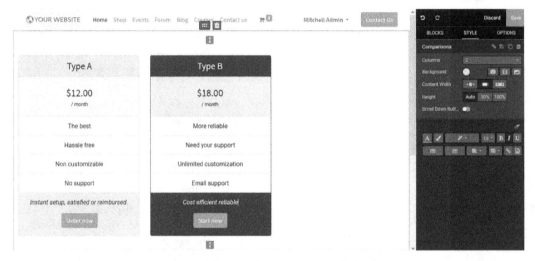

Figure 4.4 – Web page after comparison block editing

As we are clear on the comparison block tool available in the Odoo website builder in the features block, let's now jump on to the next type of block operations – the team block – in the next section.

Team block

The **Team** block of the features block available in Odoo is a tool that will be useful for you to describe your team of employees who run the company. This space can be configured in a website for the employees to review their experience working for the company or can be used to highlight the qualification and experience of the employees that make them valuable to the company. The following screenshot depicts the **Team** icon in the editing window of the website builder:

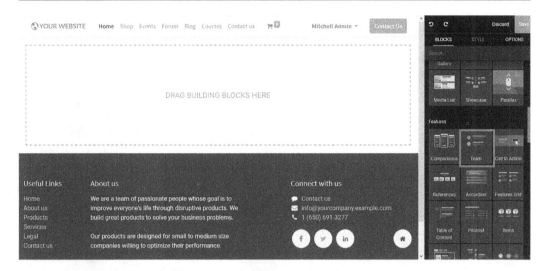

Figure 4.5 – Team block icon in the web page editing window

After choosing the team block available, you can drag and drop the default block to the desired location in the web page. Moreover, a default content and image description will be available, which can be modified. You can see the web page preview of the team block in the following screenshot:

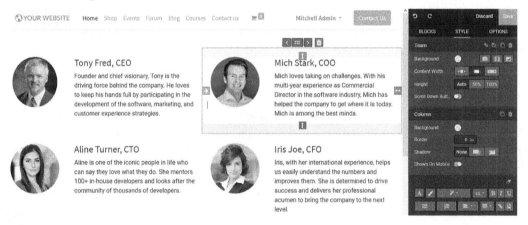

Figure 4.6 – Web page preview of the team block

The images can be changed along with the content of each employee. Moreover, as you can see in the following screenshot, there are style editing tools and options available that will help you to configure the respective block in the web page as per your requirements:

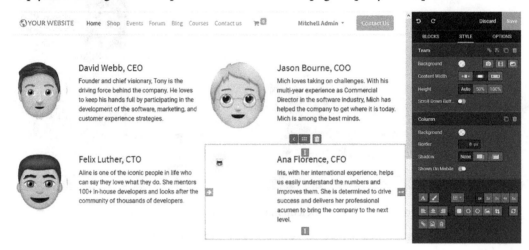

Figure 4.7 – Resultant web page after team block editing

In addition, these style tools and the options available can be configured in a similar way as done in all other block operations.

Now that we are clear on the operations of the team block, let's now jump to the call to action.

Call to action block

The **call to action** block is an attractive block design available in the Odoo website builder. It can be used to provide a better banner to the web page at the bottom or in the middle with an option to choose that would direct the visitor to a different web page. Here's how the icon looks in the editing window:

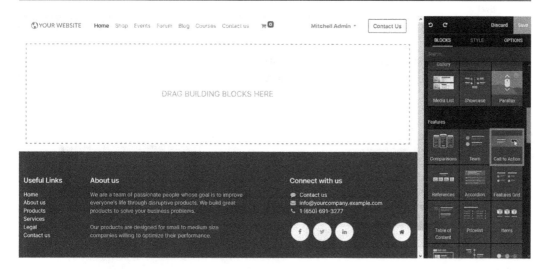

Figure 4.8 – Call to Action block icon on the web page editing window

The **Call to Action** icon can be seen under the features block section of the website editing menu. Upon choosing it, you can drag and drop the block in the desired location on the web page. Initially, the block will depict the default content available, which can be edited and modified along with the smart button option available. The following screenshot depicts the web page preview of the call to action block in a web page:

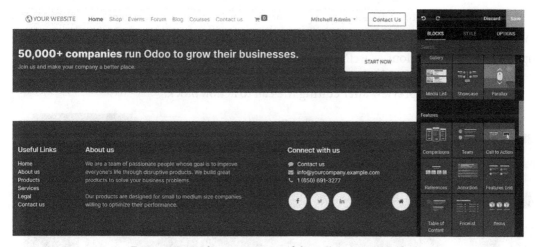

Figure 4.9 – Web page preview of the call to action block

In the smart button configuration window, the web page URL or the email can be configured along with the page anchor if available. Additionally, the type of the option along with the size and the style of the smart button can be configured. The following screenshot depicts the smart icon configuration window:

Link to   ✕

| URL or Email | /contactus | | Preview |
| Page Anchor | | ⌄ | START NOW |

Hint: Type '/' to search an existing page and '#' to link to an anchor.

Page Anchor: Loading...

Type   Link Primary ✔ Secondary

Size   Large   ⌄

Style   Default   ⌄

⟳ Open in new window

Save   Discard

Figure 4.10 – Smart icon configuration window in the call to action block

Once the block and its allocated content area are configured and described, you can now opt for the various options as well as the style configuration tools available, which will help you to configure the block according to your specification. The following screenshot depicts the web page preview after editing the call to action block:

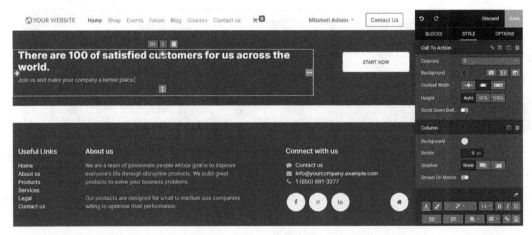

Figure 4.11 – Resultant web page after call to action block editing

In addition, the configuration style as well as the options available can be configured in a similar way as for all the other block operations.

As we are clear on the configuration of the call to action block, we can now move on to the next section, where the references block will be described.

References block

The **References** block is a useful tool in web page design, where the customer list and well-known clients of your company can be listed. Moreover, the customer logos can be used, which will provide an attractive design and add to the illustration aspects, as shown in the next screenshot:

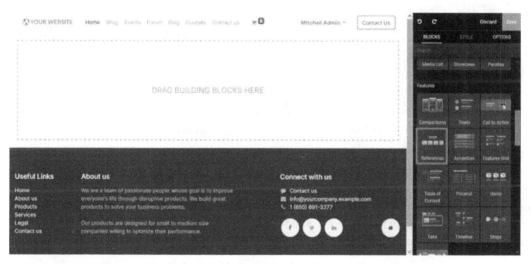

Figure 4.12 – Reference block icon in the web page editing window

The **References** block can be obtained from the editing menu under the features block description. Upon choosing it, the block can be dragged and dropped into the web page and can be further customized. The following screenshot shows the web page preview of the references block:

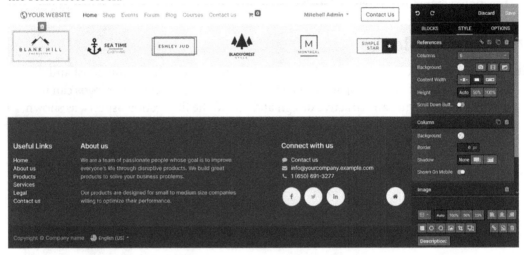

Figure 4.13 – Web page preview of the references block

The default content will be a strip of images that will depict the company logos, which can be modified by double-clicking on it and providing your clientele with their logos or company images. In addition, you can add as many reference blocks as you need to describe multiple companies. This can be used to describe certain company products as well as to provide a pictorial description of them. The following screenshot shows the web page after the references block editing is done:

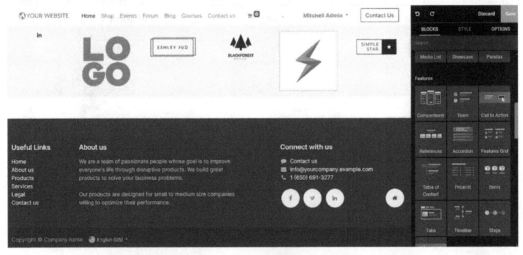

Figure 4.14 – Resultant web page after references block editing

There are further editing and style configuration tools available in all other block operations. Now let's move on to the next block under the features block – the accordion block – in the next section.

Accordion block

The **Accordion** block is a block type under the features block category and you can use it for describing certain policies of the company operation or the website. These terms and conditions can be customized and well drafted to suit your company's operation policy. The following screenshot depicts the **Accordion** icon available in the features block category:

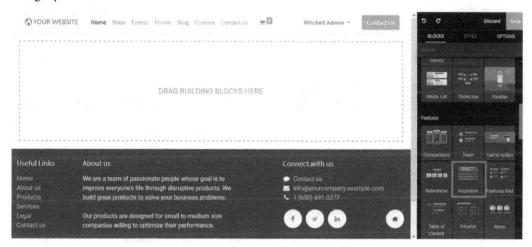

Figure 4.15 – Accordion block icon in the web page editing window

Once the accordion block is chosen, you can drag and drop it into the web page and to a distinctive location on your selected web page. Moreover, the default content available can be modified. In addition, you can add or remove a section and configure the link to the section, as shown in the following web page preview of the accordion block:

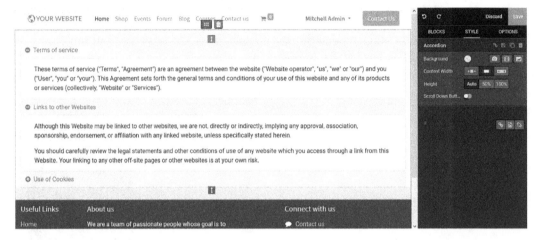

Figure 4.16 – Web page preview of the accordion block

Upon double-clicking the section heading, you will see a window where the required link on the website can be configured. Additionally, it can be configured as an email address and the type can also be defined:

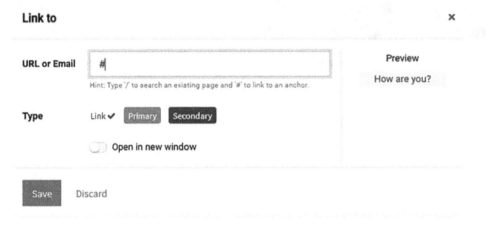

Figure 4.17 – Smart icon configuration window in the accordion block

Once the default content is edited to suit your company policy and terminology description, you can edit the style and the options available. They are similar to the ones available in all the block operations. The following screenshot shows the resultant web page after the accordion block editing:

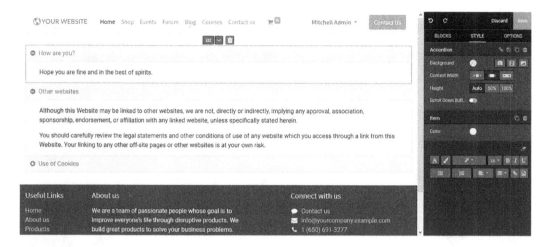

Figure 4.18 – Resultant web page after accordion block editing

As we are clear on the accordion block, let's now jump on to the next section, where the features grid block is being described.

Features grid block

The **features grid** block provides you with a space to describe the salient features of your business. Here you can list the features of products and services and describe them in a brief section. The following screenshot depicts the **Features Grid** icon in the web page editing menu:

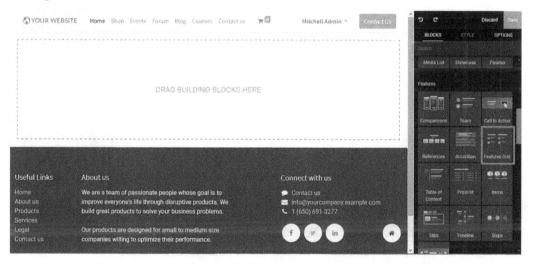

Figure 4.19 – Features grid block icon in the web page editing window

Upon selecting the features grid block from the edit menu under the features block, drag the block into the respective page allocation in your web page. Moreover, it comes with a default content of features, which helps to define the ones you require, as shown in the following screenshot:

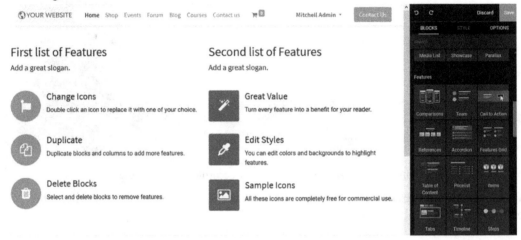

Figure 4.20 – Web page preview of the features grid block

Moreover, the icons available can be edited along with the content description tool to suit your requirements. In addition, there are numerous editing and style configuration options available, similar to the ones available in all other block operations. The following screenshot showcases the resultant web page preview after the grid block editing:

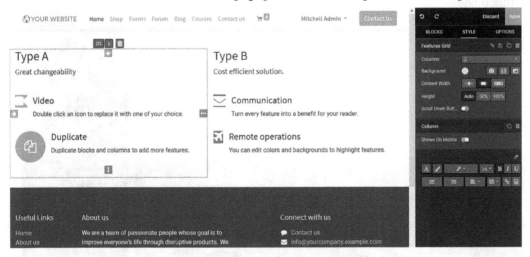

Figure 4.21 – Resultant web page after features grid block editing

Now, as we have an understanding of the features grid block, let's now jump to the next section, where the table of content block is described.

Table of Content block

A Table of Contents in a website will be a helpful tool to the readers, as they can navigate easily to the content they want. The Odoo features block category in the Odoo website builder provides you with the **table of content** block tool that will help you to describe the contents in the web page. The table of content section can also be used to describe the contents of a book and can be obtained from the editing menu of the block operations, as shown in the following screenshot:

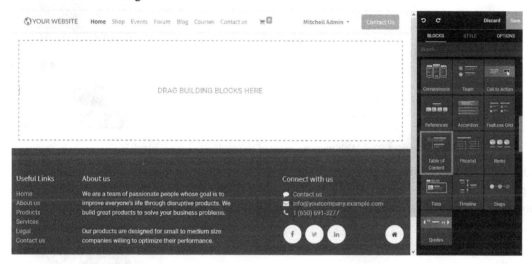

Figure 4.22 – Table of content block icon in the web page editing window

Once the table of content block is chosen, you can drag and drop it in the desired location in a web page. A default content and description will be present to describe how the operation works, as shown in the following screenshot, and you can edit it to your requirements:

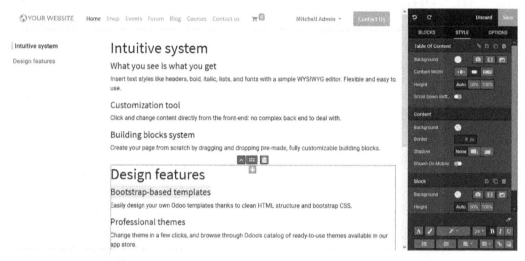

Figure 4.23 – Web page preview of the table of content block

Once the content editing is completed, there are block editing and style configuration options similar to the ones available in all other block operations of the platform. The following screenshot shows the resultant web page after the table of contents block editing:

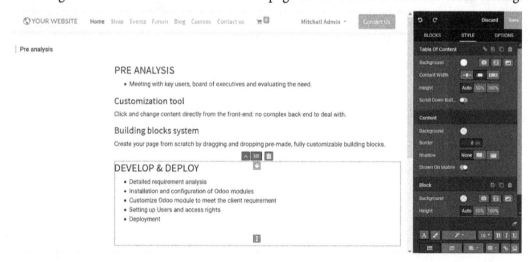

Figure 4.24 – Resultant web page after table of content block editing

As we are clear on the table of content block available in the Odoo website builder, let's now jump to the next section, where the pricelist block is explained.

Pricelist block

You may have seen the price list of a company's products on their website and may have wondered how they have configured it in such an attractive manner. The Odoo website builder has an apt block tool called the **Pricelist** block falling under the features block category. The following screenshot shows the **Pricelist** icon available in the website editing menu:

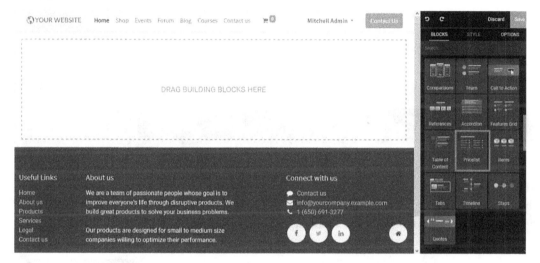

Figure 4.25 – Pricelist block icon in the web page editing window

The Pricelist block will allow you to configure the product price list attractively. Upon choosing the pricelist block, drag and drop it to the appropriate location in your web page. Initially, there will be a default content description to explain to you how to configure it. The following screenshot shows a web page preview of the pricelist block in a web page:

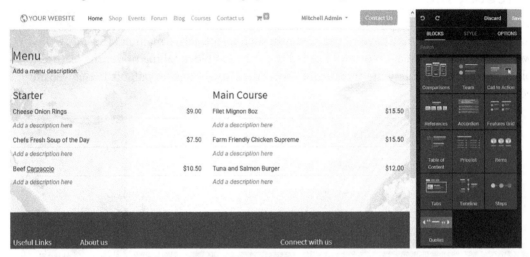

Figure 4.26 – Web page preview of the pricelist block

The contents, along with the background image, can be edited and modified. Moreover, there are style editing and optional tools to configure the block as available in all the block operations throughout the website building using the Odoo website builder. The following screenshot depicts the resultant web page with the pricelist block after editing:

Figure 4.27 – Web page after pricelist block editing

As we are clear on the pricelist block operation in website building using Odoo, now we can move on to the next block under the features block in the upcoming section.

Items block

Another interesting block tool in the Odoo website builder under the features block section is the **Items** block, which can help you to list the product with illustrative content. The **Items** icon can be chosen from the editing menu available on all the web pages and under the features block, as shown in the following screenshot:

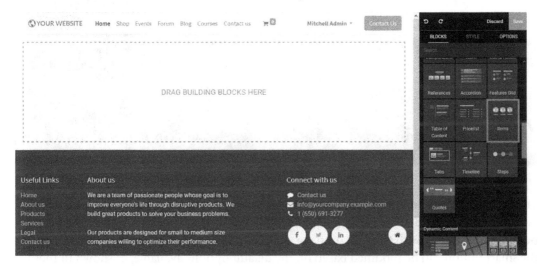

Figure 4.28 – Items block icon in the web page editing window

Upon choosing the **Items** block, you can drag and drop it to be described in the web page in an appropriate location. There will be an image block available as a default, along with the options to choose for an external link to be provided, as shown in the following screenshot:

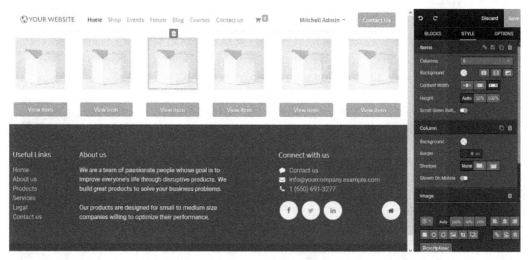

Figure 4.29 – Web page preview of the items block

The external link option under each item can be configured by selecting the smart icon available and will be depicted with the following window. Here, the URL or the email can be configured, along with the type of the option. Moreover, the size as well as the style of the smart button can be modified based on the default options available, as shown in the following screenshot:

Figure 4.30 – Smart icon configuration window in the items block

Furthermore, the image can also be edited and removed if required, and you are provided with an option to remove a section as per your requirements. There are configuration options as well as editing tools available in all the block operations in the Odoo website builder:

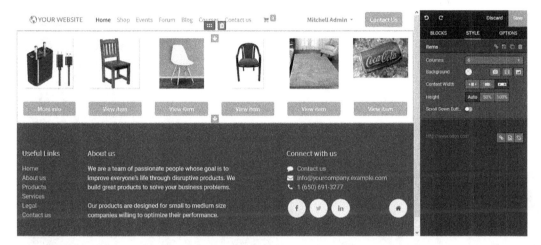

Figure 4.31 – Resultant web page after items block editing

Now, as we are clear on the items block and its allocated configuration settings, let's now move on to the next block under the category of the features block – the tabs block – in the next section.

Tabs block

There should be a navigational option that allows the visitors of a website to move around the web pages. The **Tabs** block section in the Odoo website builder will provide ample options to configure the navigational operations and can be chosen from the website editing window as shown in the following screenshot:

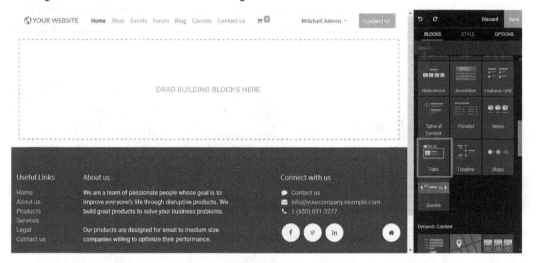

Figure 4.32 – Tabs block icon in the web page editing window

Upon choosing the tabs block, you can drag and drop it to a suitable location in the web page. Normally, they are either located at the top of the web page or at the bottom. However, it can be configured based on your needs and necessity. Moreover, there will be default content displayed upon choosing it that will provide an understanding of it, as shown in the following screenshot:

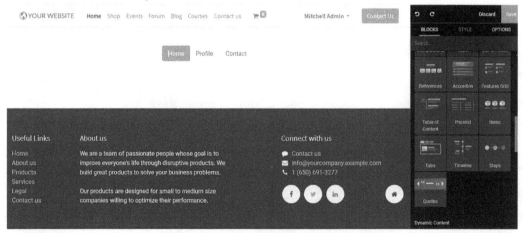

Figure 4.33 – Web page preview of the tabs block

The tab options can be configured to direct you to the YouTube allocated webpage. To do so, double-click on the icon and you will be shown the following window to configure the URL or email address. The option can be labeled and a type can be chosen. In addition, numerous tabs can be allocated:

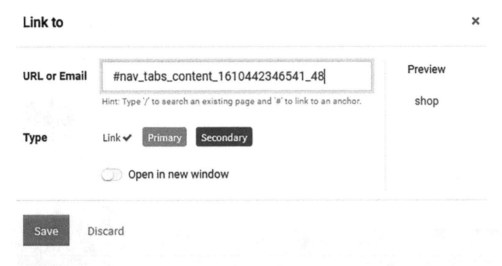

Figure 4.34 – Smart icon configuration window under the tabs block

Once the configuration is done, you can now edit the style and appearance options available in all the block operations of the platform. This will allow you to have a customized appearance for the block and its operations, as shown in the following screenshot:

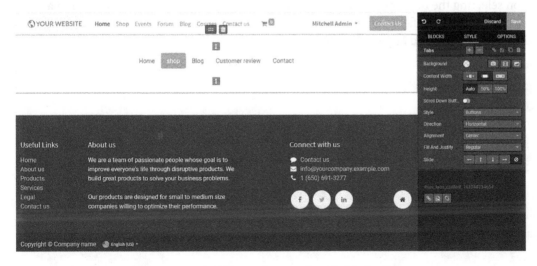

Figure 4.35 – Resultant web page after tabs block editing

Now, as you are clear on the tabs block operations in the Odoo website builder, let's move on to the next features block – the timeline block – in the next section.

Timeline block

On your company website, you may need to describe the timeline operations of your company to showcase your achievements. The Odoo **Timeline** block in the website builder will provide you with a design to describe the timeline of the company and can be chosen from the website editing window, as shown in the following screenshot:

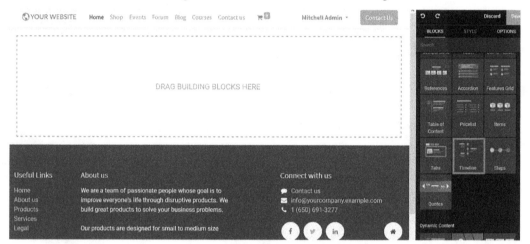

Figure 4.36 – Timeline block icon in the web page editing window

Upon selecting the timeline block, you can drag and drop it to the desired location in a web page. Initially, default content will be depicted, as shown in the following screenshot, which can be modified as per your requirements:

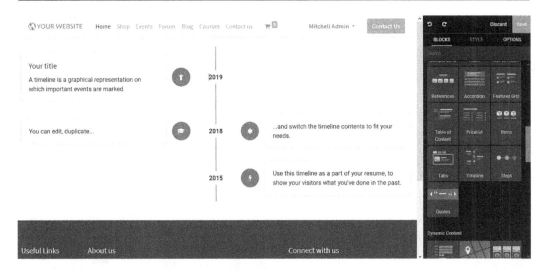

Figure 4.37 – Web page preview of the timeline block

The content, along with the illustrative image, can be modified and changed or removed. Moreover, there are distinct style configuration options available just as in every other block operation. The following screenshot shows the resultant web page after the timeline block editing:

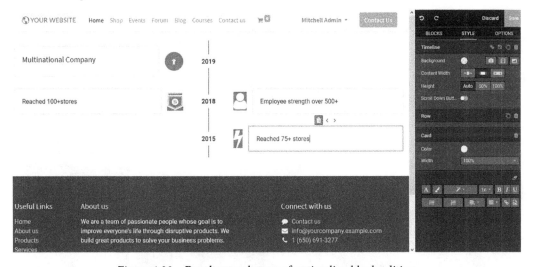

Figure 4.38 – Resultant web page after timeline block editing.

As we are clear on the Timeline block operations of the Odoo website builder, let's move on to the next features block type – the steps block – in the next section.

Steps block

The **Steps** block in the category of features blocks in the Odoo website builder application will provide you with a preconfigured design on the operations of ordering or company operations, which can be chosen from the web page editing menu, as shown in the following screenshot:

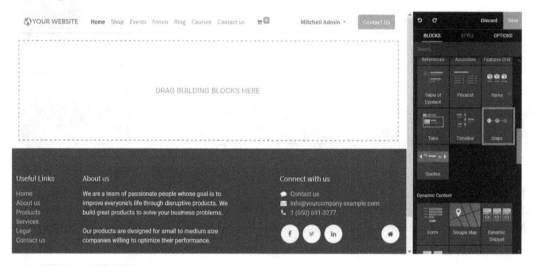

Figure 4.39 – Step block icon in the web page editing window

Once the step block is selected, you can drag and drop it into the web page at the desired location. By default, there will be content that will provide an idea of how to configure the block, as shown in the web page preview in the following screenshot:

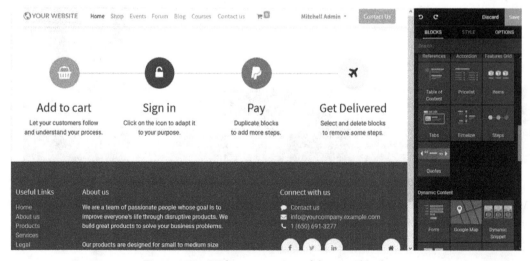

Figure 4.40 – Web page preview of the steps block

In addition, the contents available can be edited and modified along with the pictorial representation. Step columns can be added and removed. Moreover, there are editing tools and style configuration options available that will help you to configure the block design, as shown in the following screenshot:

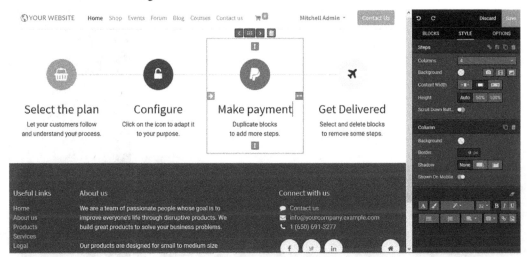

Figure 4.41 – Resultant web page after step block editing

As we are clear on the **Steps** block and its configuration operations, let's now move on to the last features block – the quotes block – in the next section.

Quotes block

The **Quotes** block is the last features block type and is the one that will provide you with the simplest of designs to provide various quotes by your employees, customers, and prominent people in your sector of business. Moreover, the **Quotes** icon can be chosen from the web page editing menu, as shown in the following screenshot:

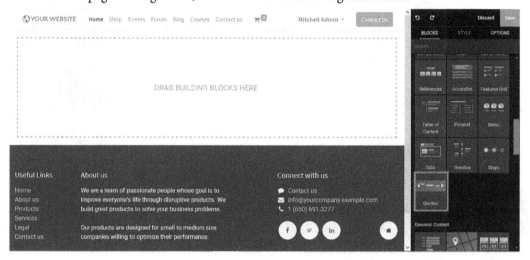

Figure 4.42 – Quotes block icon in the web page editing window

Once the **Quotes** block is selected, you can drag and drop the block in the desired location on a web page. That will initially provide you with default content that can be modified, as shown in the following screenshot. Moreover, you can add as many slides containing quotes as you want, or remove them, as seen in the following screenshot:

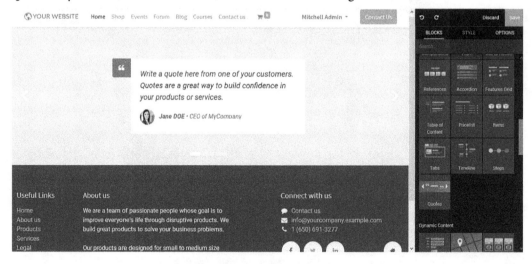

Figure 4.43 – Web page preview of the quotes block

Although you will have editing and style options to configure it, these are similar to the ones available in all other block operations. The following screenshot shows the web page preview after the quotes block editing:

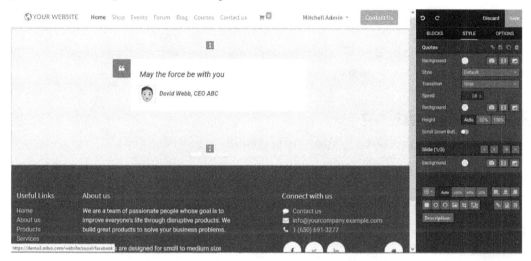

Figure 4.44 – Resultant web page after quotes block editing

In conclusion, the features block in the Odoo website builder will provide you with ample supporting blocks and types to illustrate and highlight certain aspects of the company website in brief as well as in a more descriptive way. In addition, the editing and style configuration options available will add to the appearance of the website, projecting the contents to the reader in an understandable as well as soothing manner.

Summary

In this chapter, we gained an understanding of the features block available in the Odoo website builder and how to operate and design your website using the various types of features blocks available. Now you will be able to configure the website design using the features block with ease to highlight certain aspects of your company on the website.

In the next chapter, we will learn about the dynamic content block and the various types you can use to design and configure the website according to your needs.

Questions

1. How many types of features blocks are there in the Odoo website builder?

2. Which features block can be used to provide web page navigation options in a website?

3. Which block can be used to illustrate your customers or other company logos associated with your company operations in the website?

Further reading

- *Working with Odoo* by Greg Moss, Packt Publishing
- *Learn Odoo* by Greg Moss, Packt Publishing

5
Designing a Website using Dynamic Content

You now have an understanding of the website builder and its allocated operations, especially the features block. As the Features block is used to provide details of operations, there is a requirement for other tools to describe the content of the web pages. Therefore, the Odoo website builder has a **Dynamic Content** block that will help you in defining the required contents of a website, and will be explained in detail in this chapter.

In this chapter, we will be covering the following aspects:

- Introducing Dynamic Content blocks
- Exploring the types of Dynamic Content block

By the end of this chapter, you will have the understanding and ability necessary to design a website using the dynamic content block available in the Odoo website builder.

Technical requirements

Although you might be familiar with the website building applications in Odoo, as a beginner in the field you might need an understanding of Odoo and its website builder module. Moreover, a system with Odoo installed or a database will be required to run the platform and design a website using the website builder.

Introducing Dynamic Content blocks

As mentioned in the initial chapters, there are different types of website building blocks in the Odoo website builder and, among these, the Dynamic Content block is considered the third type of available block. While a structure block is used to provide structure to the website and a features block describes the additional features of a company, a Dynamic Content block is used to provide structure to the content, which should be described in the web page.

Furthermore, with various other aspects, such as when receiving an alert or a message to the visitors of a web page, and the other dynamic block comes under the Dynamic Content block classification. You can use these operation blocks for website building and describing the content of a website in an illustrative, as well as an attractive, way.

In the range of the Dynamic Content blocks available in the Odoo website builder, there are approximately 14 types of Dynamic Content block that will help you to describe the content of your website with ease. In addition, there are various editing and style configuration options available with each block under the dynamic content blocks, which are similar to the ones available in all other block operations of the Odoo website builder. However, in this chapter, these settings and style options will not be explained; rather, you can read the section of the banner block, available in *Chapter 3, Introduction to Blocks – Structure Blocks,* to get an in-depth knowledge on it. As we have an understanding of Dynamic Content blocks, let's now move on to understanding the different types available.

Exploring the types of Dynamic Content block

The dynamic content blocks in the Odoo website builder application can be described as the ones used to define the contents of a website. Rather than structure or feature blocks, which bring structure to a web page, a dynamic content block will help you to describe the contents of a web page using various block operations available under the classification. In total, there are 16 types of dynamic content block available in the Odoo website builder. However, if you couldn't find a certain block mentioned here under the Dynamic Content blocks, make sure the Developer mode is activated:

- Form block
- Google Map block
- Dynamic Snippet block
- Dynamic Carousel block
- Dynamic Products block
- Viewed Products block
- Products Search block
- Blog Posts block
- Events block
- Newsletter block
- Newsletter Popup block
- Popup block
- Facebook block
- Countdown block
- Discussion Group block
- Twitter Scroller block

These will be explained to you from the next section onward.

Form block

The **Form** block tool is the initial block under the classification of the dynamic content block available in the Odoo website builder. Moreover, a form block is used to provide a form for the visitors to fill out, asking about the products and services your company offers. In addition, the contents of the form can be modified, removed, or added to suit your website requirements. You can select the **Form** block from the web page editing window and drag it to a desired location in the web page, as depicted in the following screenshot:

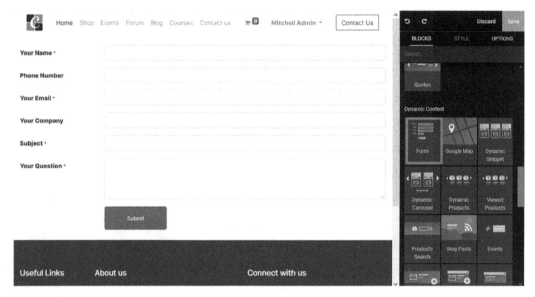

Figure 5.1 – Form block icon in the web page editing window

Once the Form block is selected from the list of dynamic content blocks available in the website builder, it can be dragged and dropped into the desired web page locations available. Moreover, it will provide you with a default content form for visitors to fill out if they need to ask a question. The contents can be edited and options in the form can be added or removed, but the basic block structure available will remain the same.

In addition, the **Submit** button or icon can be configured to direct visitors to a different web page or form set to the company's email address. This can be done in the window, as depicted in the following screenshot:

Link to   ✕

URL or Email | # |

Hint: Type '/' to search an existing page and '@' to link to an anchor.

Preview

Submit

Type   Link   [Primary ✔] [Secondary]

Size | Large   ⌄ |

Style | Default   ⌄ |

⬭ Open in new window

[Save]   Discard

Figure 5.2 – Smart icon configuration window under the form block

Once the form and the submit options are configured, there are various type and configuration options available, which are similar to all other block operations available in the Odoo website builder. Furthermore, as shown in the following screenshot, there are various block configuration options to provide style to the block on its own:

Figure 5.3 – Web page after form block editing

As we have an understanding of the form block available in the Odoo website builder, let's now move on to the Google Map block in the next section.

Google Map block

Have you ever considered integrating Google maps into your company's website to help visitors navigate to the facility or store with ease? The Odoo website builder has a **Google Map** block available that will help you to integrate Google Maps in to your web page! The following screenshot shows what the **Google Map** block looks like in the website editing window:

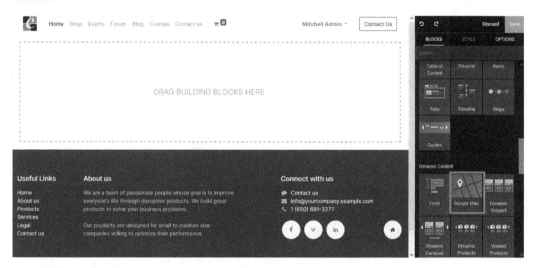

Figure 5.4 – Google Map block icon in the web page editing window

Upon selecting the **Google Map** block (available as the second block option under **Dynamic Content**), you can drag and drop it into the web page, upon which you will be presented with the following window to configure the **Google Map API** key:

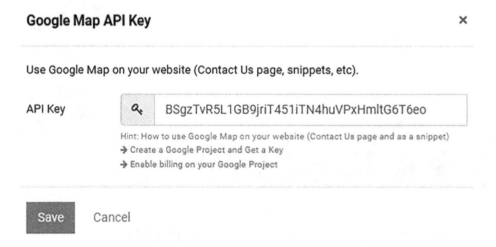

Figure 5.5 – Google Map block API configuration window

In addition, there are external link options in the configuration window that will direct you to the Google **Project Billing** window. Moreover, the Google project (such as **Google Map** integration) can be paid for, based on subscription charges, to avail a functionality, you can get started from the window, as depicted in the following screenshot:

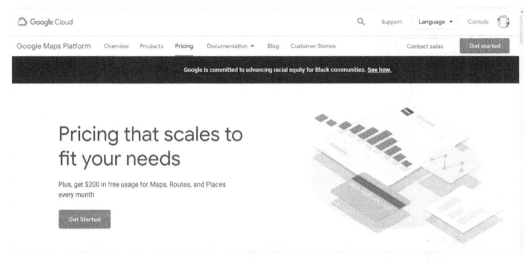

Figure 5.6 – Google Cloud web page

Furthermore, if you need to create and generate an API key, you can register the **Maps JavaScript API** or **Maps Static API** application in **Google API Console** using the **Create Google project** external link and get a key option available that can be configured in the web page, as depicted in the following screenshot:

Figure 5.7 – Google Map block API configuration web page

There are no further editing and style configuration options available with this block tool. As we are clear about the Google Map block available under the Dynamic Content block, let's now move on to a new block: the Dynamic Snippet block.

Dynamic Snippet block

The **Dynamic Snippet** block is the third block available under the classification of dynamic content blocks in the Odoo website builder. The Dynamic Snippet block will help you to provide a snippet of operations to describe the contents of a web page. Here's what the icon looks like in the web page editing window:

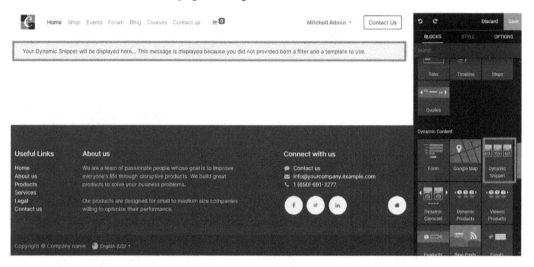

Figure 5.8 – Dynamic Snippet block icon in the web page editing window

Upon choosing the **Dynamic Snippet** block, drag and drop it to the desired web page location. Moreover, default content will be displayed to you to configure a filter and template to be used to display the content. Upon choosing the **Style** menu from the existing window available on the right-hand side of the screen, you can configure a template and filter to display the content.

Furthermore, there are only two default filtering options as well as template options available for you to choose from. Additionally, you can add various filters and templates, using the backend of the platform and technical knowledge in coding to do so. In addition, you can only add 16 elements to the web page using this block tool, and the display option of the items in the devices can be configured for normal devices such as laptops, desktops, and other gadgets with big screens, as well as small devices such as mobiles, tablets, and other smart gadgets.

The following screenshot depicts the web page as if the filtering had been done based on country, and a **Header Image Footer Card** template has been chosen:

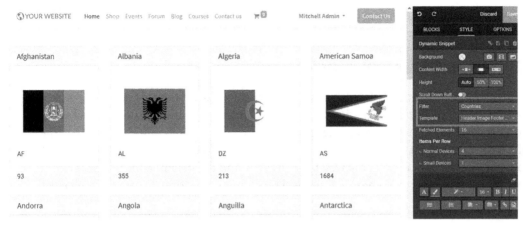

Figure 5.9 – Web page preview of the Dynamic Snippet block with the Countries filter and Header Image Footer Card template

Now, in the following screenshot, a **Products** filter is chosen, and an **Image title footer card** template is selected. The products described in the company web page or the e-commerce platform can be listed out on other web pages by using this block tool. Additionally, items to be displayed in each device can also be configured as per for the other template and filter described previously:

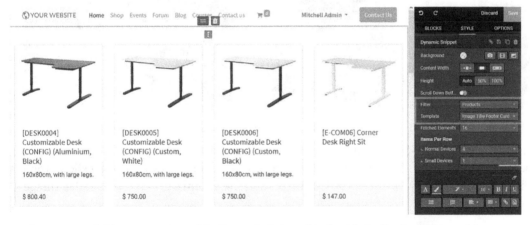

Figure 5.10 – Web page preview of the Dynamic Snippet block with the Products filter and Image Title Footer Card template

Moreover, there are various style and configuration options available to the platform that are similar to the ones available in all other block operations in the Odoo website builder.

As we have an understanding of Dynamic Snippet blocks, let's now move on to the next section, where the Dynamic Carousel block is described.

Dynamic Carousel block

This is another Dynamic Content block tool available. The **Dynamic Carousel** block will provide you with a carousel section in the web page that has slides to display and can be auto-scrolled or moved according to the website visitor's preferences. Moreover, the block can be dragged and dropped on to any web page based on your requirements. The following screenshot depicts the web page preview and the default editing information for you to configure the **Dynamic Carousel** block upon selecting it from the web page editing menu:

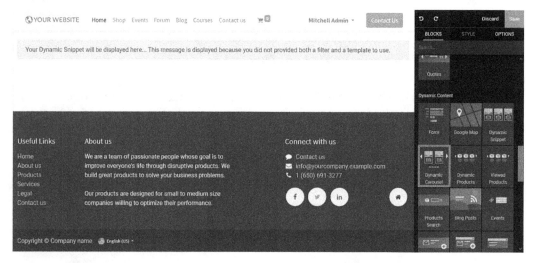

Figure 5.11 – Dynamic Carousel block icon in the web page editing window

As with the Dynamic Snippet block, the Dynamic Carousel block can be configured based on the filtering and template options available. The following screenshot shows the **Filter** option of the block chosen as **Countries** and the **Template** option as **Header Image Footer Card**:

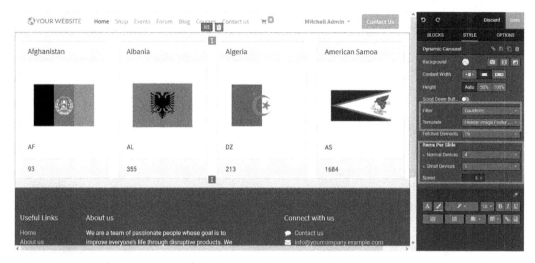

Figure 5.12 – Web page preview of the Dynamic Carousel block with the Countries filter and Header Image Footer Card template

Furthermore, there are **Items Per Slide** configuration options available that only allow you to configure four or fewer items in normal devices, and in smaller devices, you can only configure up to three items. In addition, the **Speed** option of the carousel motion with which the slides move can be set as per your requirements. These options are illustrated in the following screenshot:

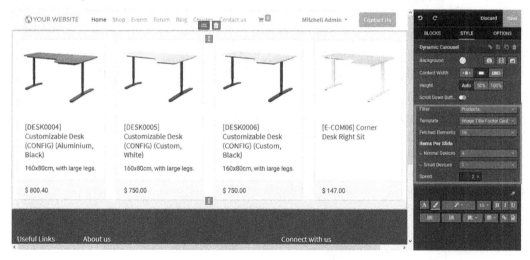

Figure 5.13 – Web page preview of the Dynamic Carousel block with the Products filter and Image Title Footer Card template

Additionally, there are various style and editing options available that are similar to the ones in all other block operations of the platform.

As we are clear about the Dynamic Carousel block, let's now move on to the next type of dynamic content block: the Dynamic Products block.

Dynamic Products block

This block type will help you to describe the product of your company or store on your website. Upon selecting the **Dynamic Products** option, you can drag and drop it onto the web page, and you will be presented with a default editing note to configure the block, as depicted in the following screenshot:

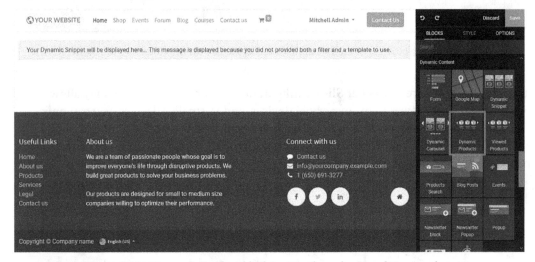

Figure 5.14 – Dynamic Product block icon in the web page editing window

Once the block is selected, you should configure the filter along with the **Product Category** to be displayed initially. Moreover, the product category defined in the backend of the Odoo website builder will be auto-depicted here, and you can choose a respective one to be displayed in this block. In addition, the filter for the display of the products can be either chosen as **Header Image Footer Card**, as shown in the following screenshot, or as **Image Title Footer Card**:

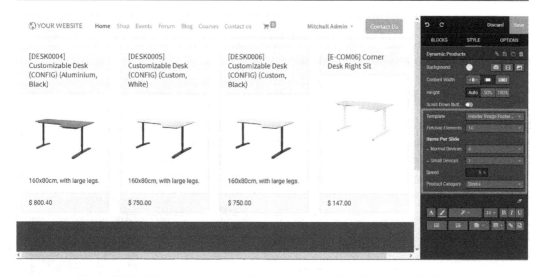

Figure 5.15 – Web page preview of the Dynamic Product block with the Header
Image Footer Card template

The operation of the **Image Title Footer Card** template is depicted in the following
screenshot. Furthermore, the total number of elements that can be displayed or fetched in
this block is 16, and the items configured to be operations in normal devices and smaller
devices can be configured based on the default numbers available. Moreover, like this,
a carousel type of block automatically moves the slides, the speed of which, can also be
configured:

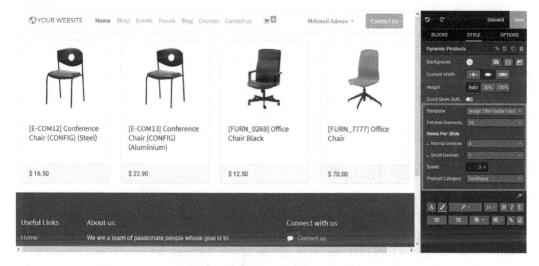

Figure 5.16 – Web page preview of the Dynamic Product block
with the Image Title Footer Card template

Additionally, there are configuration, styling, and editing options available, just as in all other block operations of the Odoo website builder. As we are clear about the Dynamic Product block, let's now move on to the next block.

Viewed Products blocks

This is a type of block operation in Odoo that comes under the classification of Dynamic Content blocks and helps you to describe the **viewed product** to website visitors. Moreover, the block can be obtained from the web page editing window under the **Dynamic Content** block section, as depicted in the following screenshot:

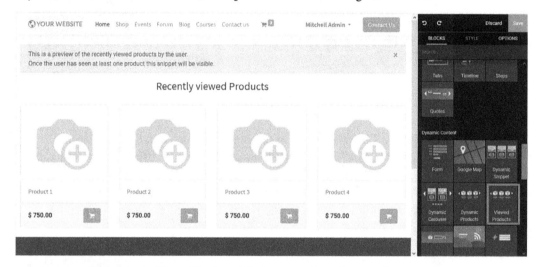

Figure 5.17 – Viewed Products block icon in the web page editing window

All products recently viewed by the visitor will be displayed to them when the block is up and running. Furthermore, there are various editing and style configuration tools available that will provide you with various customizable options to the block, just as in all other block operations in the Odoo website builder. As we now have an understanding of the Viewed Products block, let's now move on to the next block section.

Products Search block

Searching for a company product by visitors to a website is quite common in operations today. As you visit various websites, you can see the search option configurable and customized, to be displayed on all areas of the website. The Odoo website builder has a **Products Search** block tool that will help you to define the search menu of the product on your website. Upon selecting the **Products Search** block, drag and drop it to the desired location on the web page, and you will be provided with default content and a search bar, as shown in the following screenshot:

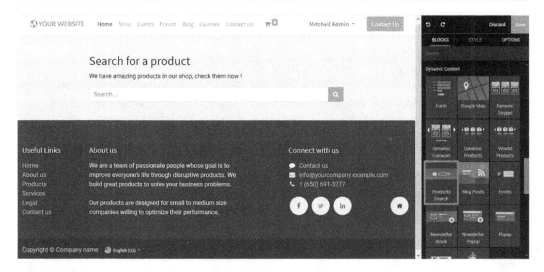

Figure 5.18 – Products Search block icon in the web page editing window

The contents available can be edited as per your requirements, and the search option can also be configured based on your requirements and your website design. In addition, there are various styling and configuration options available for the block, just as with all other block operations of the platform. This is what the web page will look like after applying **Products Search** block editing:

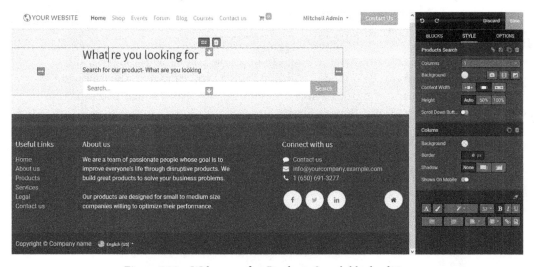

Figure 5.19 – Web page after Products Search block editing

As we are clear about the products search block in the Odoo website builder, let's now move on to the next type of block coming under the category of dynamic block content: the Blog Posts block.

Blog Posts block

This website builder block tool allows you to create a blog posting area in a web page. Moreover, the blog post will provide a structure to the blog posting being done on the web page, based on your requirements. You can choose the **Blog Posts** block from the web page editing window, and default content will be depicted, as shown in the following screenshot:

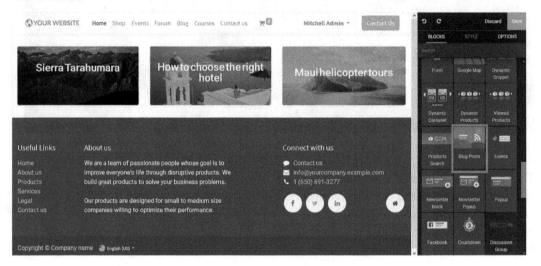

Figure 5.20 – Blog Posts block icon in the web page editing window

The default content, as well as the allocated images, can be edited and modified as per your requirements. Furthermore, a blog can be chosen along with the post, and a layout for the web page can be provided. Additionally, as you can see in the following screenshot, there are various block **Layout** options available, from which you can choose an appropriate one as per your requirements:

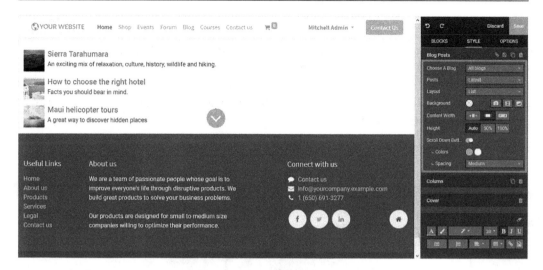

Figure 5.21 – Web page after Blog Posts block editing

Moreover, there are various configuration options as well as editing tools available similar to the ones available in all other blocks of the Odoo website builder for you to set up the block as per your design. As we are clear about the blog posts block of the Odoo website builder, let's now move on to the next block under the dynamic content block: the Events block.

Events block

You will want to list company events to visitors as this can be the best promotional tool as well as the best way to create various business opportunities. The Odoo website builder has an **Events** block that will help you to list out company events on your website. You can choose the block from the existing window and drag and drop it onto the web page as depicted in the following screenshot:

Figure 5.22 – Events block icon in the web page editing window

If the block is placed in the web page, it will consist of default content and column numbers that can be modified. Further, there are various editing as well as configuration options available that will help you to design the block as per your standards and company requirements, as can be seen in the following screenshot:

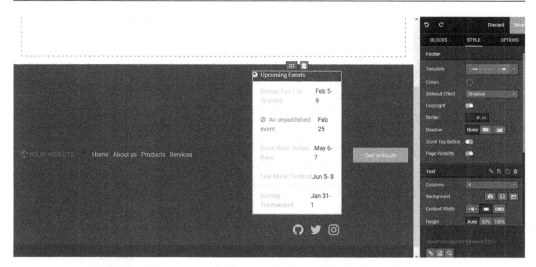

Figure 5.23 – Web page after Events block editing

Moreover, the editing and style configuration options are similar to the ones available in all other block operations of the Odoo website builder.

As we are clear about the **Events** block of the Odoo website builder, let's now move on to understanding the next block under the Dynamic Content block: the Newsletter block tool.

Newsletter Block

Online newsletters can be a better promotional and marketing tool for companies. Therefore, major enterprises will regularly publish newsletters that are based on their product services and technology associated with it and similar other competitors. The Odoo website builder has a **Newsletter block** tool that can be used to provide space in the website for visitors to subscribe to your newsletters.

Upon choosing the block from the web page editing menu, you can drag and drop it into the respective locations in the web page as required, as depicted in the following screenshot:

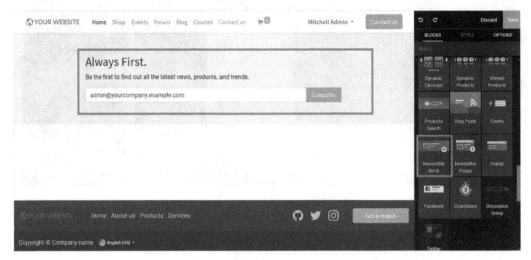

Figure 5.24 – Newsletter Block icon in the web page editing window

Initially, there will be default content provided to you that you can modify, while sticking to the standardization of the block. In addition, there are options to configure the text content of the block and a **Change Newsletter** option for subscriptions, as illustrated in the following screenshot:

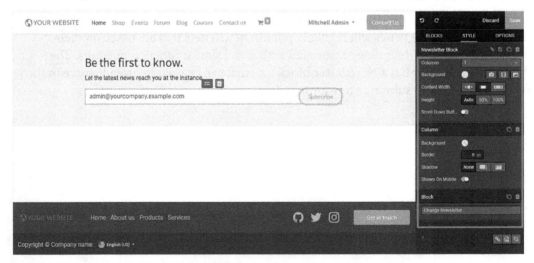

Figure 5.25 – Web page after Newsletter Block editing

The **Select Newsletter** option works when you have multiple newsletters at your disposal and you want to configure a specific one for the web page. Moreover, you can select the **Change Newsletter** option, as depicted in the previous screenshot, and you'll be provided with a pop-up window, as shown in the following screenshot, where you can provide the respective newsletter from a list of selections. Furthermore, newsletters for the subscription can be created and defined in the backend of the platform:

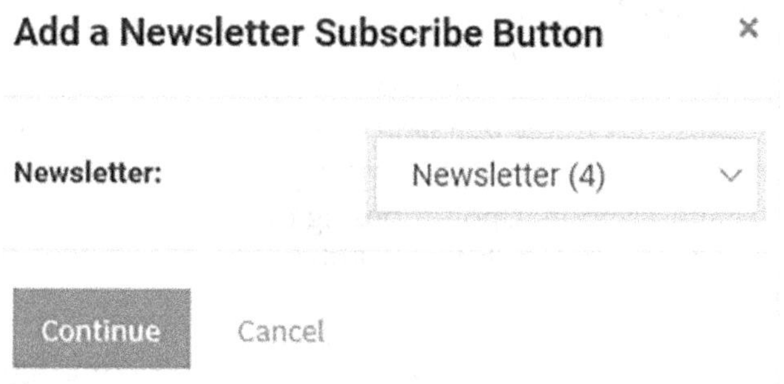

Add a Newsletter Subscribe Button   ✕

Newsletter: Newsletter (4)   ⌄

Continue   Cancel

Figure 5.26 – Pop-up window to choose a newsletter

Additionally, the **Select** option or the **Subscribe** option can be configured based on your requirements by double-clicking on the one available on the web page, and a menu will be shown, as depicted in the following screenshot. Here, the **URL or Email**, **Type**, **Size**, and **Style** options of the button can be configured:

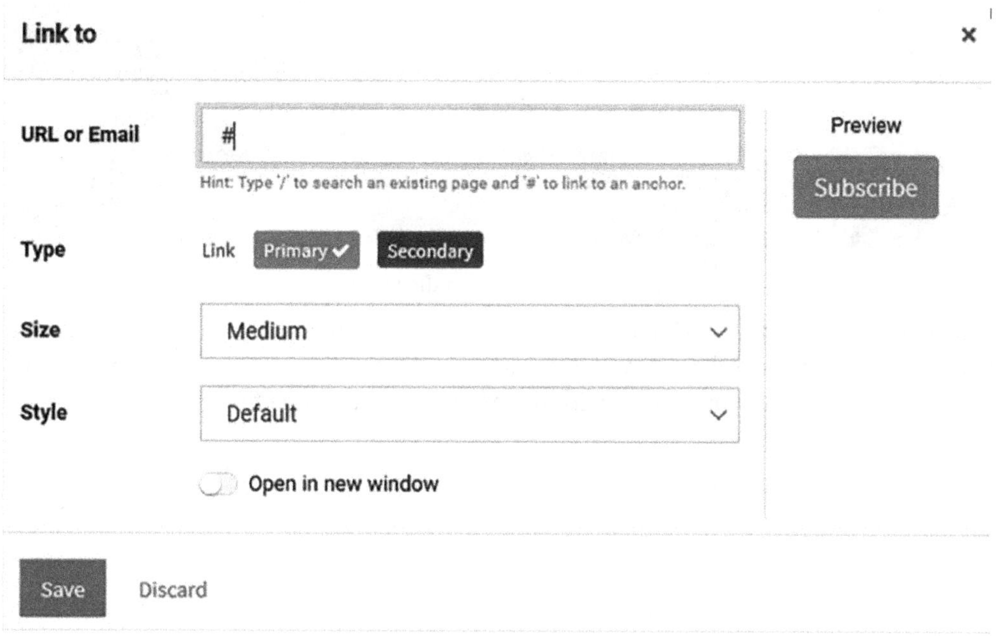

Link to   ✕

URL or Email   #

Hint: Type '/' to search an existing page and '#' to link to an anchor.

Preview

Subscribe

Type   Link   Primary ✔   Secondary

Size   Medium   ⌄

Style   Default   ⌄

⬭ Open in new window

Save   Discard

Figure 5.27 – Smart icon configuration window under the Newsletter Block

In addition, there are various configuration options as well as style configuration tools available that will allow you to showcase the block as per your requirements. Moreover, these options are similar to the ones available in all other blocks of the Odoo website builder. As we are clear about the newsletter block tool, let's now move on to the next block under Dynamic Content: the Newsletter Popup block.

Newsletter Popup block

Another way of advertising the newsletter of a company is to display a popup to visitors that can allow them to subscribe to the newsletter. The Odoo website builder has a **Newsletter Popup** tool that will provide you with a pop-up window configuration block for the newsletter subscription. Upon choosing the block from the web page editing window, you can drag and drop it to the desired location, as shown in the following screenshot:

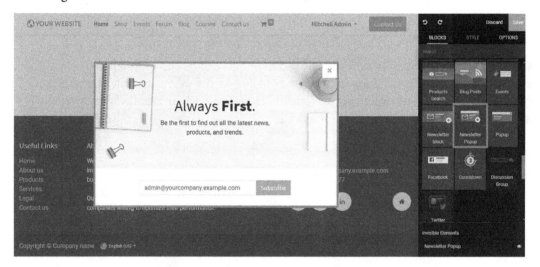

Figure 5.28 – Newsletter Popup block icon in the web page editing window

Moreover, you can configure which newsletter should be enabled for a subscription if your company has multiple newsletters, and each can be configured for different web pages. In the following screenshot, you can see with the newsletter selection window:

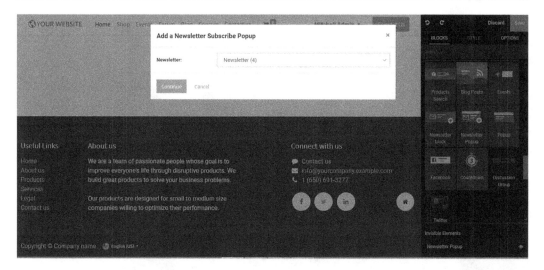

Figure 5.29 – Pop-up window to choose a newsletter

Furthermore, you can configure the **Subscription** button available by double-clicking on it. You can configure the **URL or Email** address along with the **Type**, **Size**, and **Style** options of the button, as illustrated in the following screenshot:

Figure 5.30 – Smart icon configuration window under the Newsletter Popup block

Additionally, there are editing options for the block tool, as shown in the following screenshot, whereby you can configure the **Background**, **Image**, **Position**, **Filter**, and other **Parallax** and display options as per your needs:

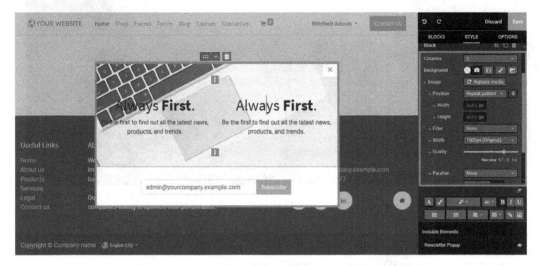

Figure 5.31 – Web page after Newsletter Popup block editing

In addition, there are various other styling as well as configuration options available with the block, just as in all other block operations of the platform. Let's now move on to another block under the Dynamic Content label: the Popup block.

Popup block

Popup messages on a website are one of the best marketing and promotional tools in order to improve business. The Odoo website builder has a **Popup** block tool that you can utilize to describe the pop-up message content of your website. Moreover, you can select the **Popup** block for the editing window and drag and drop it in the desired location in a web page, as shown in the next screenshot:

Figure 5.32 – Popup block icon in the web page editing window

In addition, you will initially be displayed with default content to provide an overview of operations that can be further modified. Additionally, the smart icon available can be configured based on your needs, which will be very useful in directing visitors to the desired web page or the email ID in order to send a direct mail to the web page. Check out the smart icon editing window shown in the following screenshot:

Figure 5.33 – Smart icon configuration window under the Popup block

Furthermore, there are options to configure the **Position, Size, Close Button Color**, and **Delay** options of the pop-up message to be displayed upon the website being visited, and many more options that can be selected to configure the pop-up window, as shown in the following screenshot:

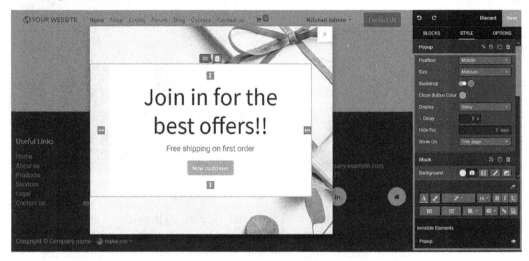

Figure 5.34 – Web page after Popup block editing

Additionally, there are various style and configuration options available that can be used to describe the popup as per your website requirements, which are similar to all the existing options available in the other block tools of the Odoo website builder. As we now have an understanding of the popup block, let's move on to the next section, where the Facebook block tool under the Dynamic Content blocks will be explained.

Facebook block

Facebook has revolutionized communication for people all over the world, and it has further served as a way for various business organizations to run their promotional as well as marketing functions. Integrating Facebook with your company website will be beneficial for your business. Therefore, the Odoo website builder has a **Facebook** block that helps you to show your company's Facebook page on your website.

You can select the **Facebook** block from the website editing window and drag and drop it onto the respective web page, as shown in the following screenshot:

Figure 5.35 – Facebook block icon in the web page editing window

You can describe your company's Facebook page in the block by providing its link in the style editing menu of the block in the website editor, as depicted in the following screenshot. Furthermore, there are options to configure the **Cover Photo**, **Timeline**, **Events**, **Messages**, **Small Header**, and **Friends' Faces** options, plus many more, based on your Facebook page:

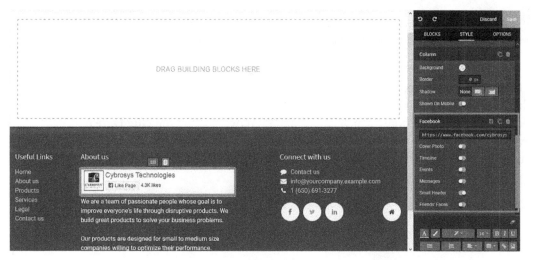

Figure 5.36 – Web page after Facebook block editing

Additionally, there are various configuration and style editing options available, just as in all other blocks of Odoo. As you now have an understanding of the Facebook block in the Odoo website builder, let's move on to the next section, where the next Dynamic Content block will be described to you: the countdown block.

Countdown block

A website countdown is used to unveil certain products or company offers. Moreover, it's an efficient marketing as well as a promotional tool for establishments. The Odoo website builder has a **Countdown** block that will help you to define the countdown for certain aspects with ease.

You can select the **Countdown** block from the web page editing window, as shown in the following screenshot, and drag and drop it to a desired location in the web page:

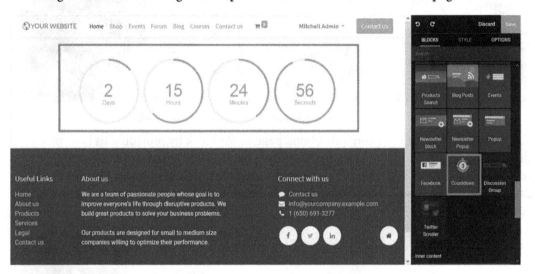

Figure 5.37 – Countdown block icon in the web page editing window

There will be initial content displayed to provide you with an insight on how to configure the block. Upon moving to the editing option, the first option to configure will be the deadline of the count done, which can be achieved by selecting the **Due Date** option available. Further, it will provide you with a calendar that you can configure with the dates and time of the due date, as illustrated in the following screenshot:

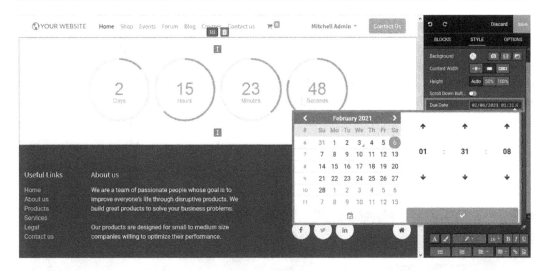

Figure 5.38 – Pop-up calendar window to choose the due date of the Countdown block

In addition, there are various other configuration options, such as providing a **URL** for the visitor after the countdown, configuring the size of the **Block, Display, Layout, Layout Background, Progress Bar**, and much more, for you to customize as per your requirements. These options are shown in the following screenshot:

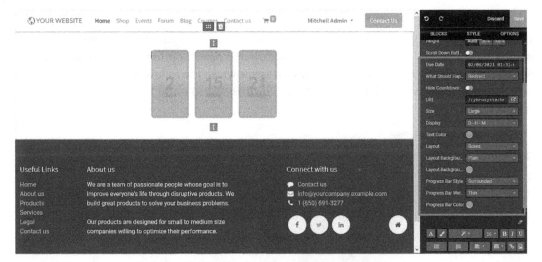

Figure 5.39 – Web page after Countdown block editing

Additionally, there are various configuration as well as styling options available that will help you to tailor the block to your requirements. Moreover, these options are similar to the ones available in all other block operations of the Odoo platform. Now, let's move on to the next block under Dynamic Content: the Discussion Group block.

Discussion Group block

Discussion groups have become an inevitable part of websites. Therefore, the Odoo platform has designed a website building block to create discussion groups. The **Discussion Group** block can be selected from the web page editing window, as shown in the following screenshot. You can drag and drop it to the desired location in the web page:

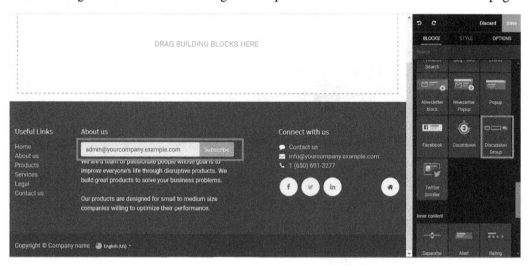

Figure 5.40 – Discussion Group block icon in the web page editing window

A default block setting will be displayed to you that can be configured upon your requirements. Furthermore, a **Subscription** option or a **Smart** option can be configured by selecting it, and a window will be displayed, as shown in the following screenshot. Here, you can configure the **URL** for visitors to be directed to or for an **Email** to be sent to a respective ID:

Link to   ✕

URL or Email   [   ]   Preview

Hint: Type '/' to search an existing page and '#' to link to an anchor.

Type   Link ✔ Primary Secondary

⊂⊃ Open in new window

Save   Discard

Figure 5.41 – Smart icon configuration window under the Discussion Group block

If you want to configure the discussion group for the web page in the case where your company has several of these, this can also be done by selecting the + option available under the **Discussion Group** from the editing menu, as illustrated in the following screenshot:

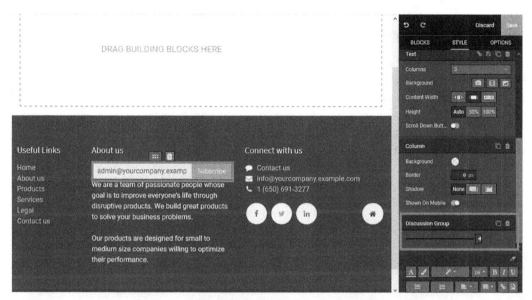

Figure 5.42 – Web page after Discussion Group block editing

Upon choosing to configure a discussion group for the web page, you will be presented with the following pop-up window where the **New Mail Channel** discussion group can be configured:

New Mail Channel   ✕

Name:

Continue   Cancel

Figure 5.43 – Discussion channel configuration pop-up window

In addition, there are various style and configuration options available for the **Discussion Group** block, as with all other blocks in the Odoo website builder. You now have an understanding of the **Discussion Group** block, so let's move onto the last block under the Dynamic Content block category: the **Twitter Scroller** block.

Twitter Scroller block

Twitter is one of the best social media platforms available today. Rather than a personal platform, it can be used as one of the best promotional tools for the marketing functions of a company. The Odoo website builder has a **Twitter Scroller** block tool to configure your company's Twitter accounts with the website of the company being designed with the Odoo website builder. Moreover, as with the other blocks, you can select the block from the website editing window and drag and drop it to the desired location in the web page, as illustrated in the following screenshot:

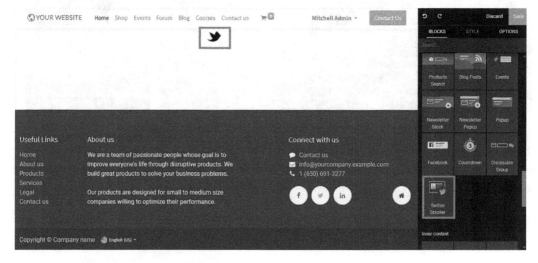

Figure 5.44 – Twitter Scroller block icon in the web page editing window

This block will be provided with default content, initially indicating to you to configure the Twitter account of your company with the Odoo website, as shown in the following screenshot. In addition, there are various other configuration and style editing options available that are similar to the ones in all other Odoo website builder block operations:

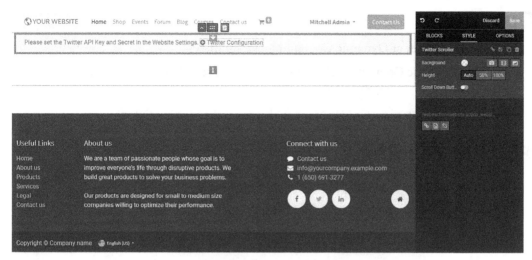

Figure 5.45 – Twitter Scroller block configuration in the web page

The Twitter company account configuration can be done with the website in the backend of the platform and under the **Company Settings** menu.

The dynamic content block tools will provide your website a space allocation to display the various content relevant to the company. Moreover, these blocks bring in additional functionalities to the operations of your web pages. In addition, various integrational block tools for social media platforms such as Twitter and Facebook will be highly beneficial in terms of the marketing and promotional aspects of the company. In general, all the blocks under the Dynamic Content block category will allow you to enhance the content of your web pages in an attractive and specific manner.

Summary

In this chapter, we gained an understanding of the Dynamic Content block under the Odoo website builder. Moreover, you now have an understanding of how to configure various different types of content of a website using **Dynamic Content** blocks. Furthermore, you have been provided with the ability to integrate external platforms such as the social media and newsletters of your company, and much more, with a web page for the visitors to access. Additionally, you now have an idea as to how the products and services of your company can be displayed on a web page using the various configuration blocks available under the Dynamic Content blocks of the Odoo website builder.

In the next chapter, we will be playing with Inner Content block tools available from the Odoo website builder.

Questions

1. How many types of Dynamic Content blocks are there?

2. Which Dynamic Content blocks help you with the social media aspects of the website?

3. How can you integrate Google Maps with your website using Odoo?

Further reading

- *Working with Odoo* by Greg Moss, Packt Publishing
- *Learn Odoo* by Greg Moss, Packt Publishing

6
Inner Content Block Tools

In the preceding chapters, we dived into the essential aspects of website building using Odoo. Moreover, in the previous chapter, we focused on the Dynamic Content block tools available to design your web page contents. This chapter will provide you with an insight on further content configuration tools available in the Odoo website builder, called **Inner Content blocks**. In this chapter, we will be covering the following topics:

- Understanding inner content blocks
- Exploring the types of inner content block

By the end of this chapter, you will have the ability to design a web page using the inner content blocks available in the Odoo website builder.

Technical requirements

If you have been reading from the beginning of the book, you will have an insight into the website building aspects of Odoo. However, for a beginner, you should have basic knowledge of the Odoo website builder, website building, and using the Odoo platform. Moreover, you will require a system with Odoo installed on it and in perfect operating condition.

Understanding Inner Content blocks

The previous chapters have been concentrating on providing a structure in design to the website which is only the initial aspect required in website building. The **Inner Content** blocks available in the Odoo website builder will provide definite tools to provide a structure and design for the contents available in the web page. With tool blocks to provide ratings, product searches, newsletters, and many more, the inner content block will provide a structure to the content description in the web page.

Content in a website is of utmost importance, along with the presentation. While website building in Odoo, you can define any form of content you wish and it is entirely up to you to define whatever is necessary for the company website. Moreover, the Odoo website builder will provide tools such as inner content blocks that will act as the descriptive medium where the contents of the website can be defined.

There are, in all, 13 types of inner content block classification forms that you can choose to provide a design to the website content. Moreover, this will provide numerous tools for style editing as well as configuration tools and functions to define the blocks as per your requirements and web page design. Now let's move on to understanding the types of inner content blocks available in the Odoo website builder.

Exploring the types of inner content blocks

There are various block operations available in the Odoo website builder and they have been described in the previous chapters. Of these, the **Inner Content** block is the last one of the classifications. In the inner content block, there are exactly 13 types of blocks being defined, which will help to provide you with definite designs for the content on the web page.

Here is the list of the inner content blocks available in the Odoo website builder:

- Separator block
- Alert block
- Rating block
- Card block
- Share block
- Product search block
- Newsletter block
- Text highlight block

- Chart block
- Progress bar block
- Badge block
- Blockquote block
- Speaker bio block

Furthermore, there are configuration and styling options available for each block operation, similar to the ones available in all other block tools. Moreover, you can read the *Banner block* section in *Chapter 3, Introduction to Blocks – Structure Blocks,* to gain an in-depth knowledge of it. Let's now move on to understanding the various types of blocks under the classification of inner content blocks in detail.

Separator block

The **Separator** block is the first type of block under the classification of inner content blocks in the Odoo website builder. This block provides an illustrative line in the web page in the description content. This can also be used as an illustrative aspect to separate the contents available for a much clearer presentation. You can access the Separator block from the web page editing menu and drag and drop it in the desired location on the web page, as depicted in the following screenshot:

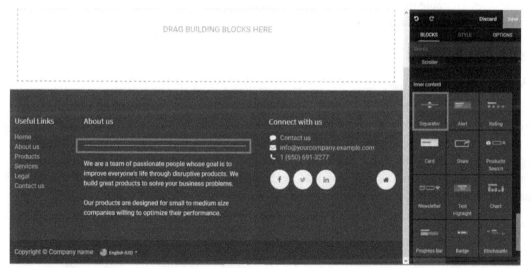

Figure 6.1 – Separator block icon in the web page editing window

Additionally, there are various configuration options available to edit the illustrative separator line, such as the **Border**, **Width**, **Size**, and **Color** editor options. These can be easily configured and obtained from the style configuration menu, as shown in the following screenshot:

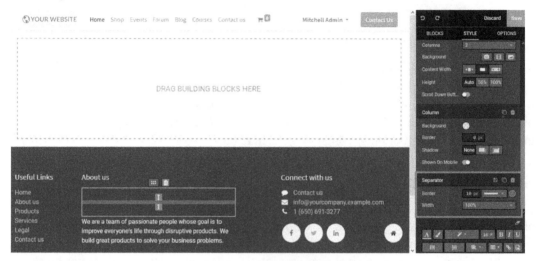

Figure 6.2 – Web page after separator block editing

Moreover, there are various style and configuration options available, which are similar to the ones available in all other block operation tools of the Odoo website builder. Now you should be clear about the separator block, therefore let's move on to the next section, where the alert block under the inner content block types will be explained.

Alert block

The **Alert** block available in the Odoo website builder allows you to create an alert message on the website if an operation is done inappropriately. Moreover, this is not a pop-up alert message as seen in the various other websites. Rather, it serves as a warning content by which the company can alert the visitors. You can choose the Alert block from the website editing window and drag it to be dropped in the desired location on the web page, as shown in the following screenshot:

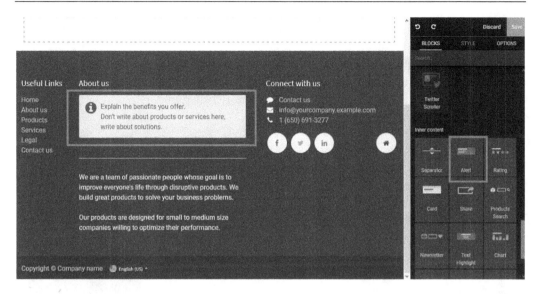

Figure 6.3 – Alert block icon in the web page editing window

Moreover, there are various editing and configuration tools available in the alert block editing window, as depicted in the following screenshot, such as **Type**, **Width**, **Size**, **Color**, and **Alignment** of the block, and the message depicted can be modified. Furthermore, the contents can also be edited as per your requirements:

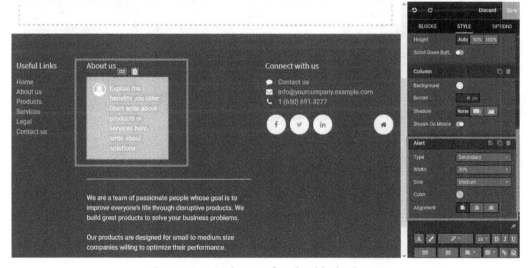

Figure 6.4 – Web page after alert block editing

The Alert block is one of the best ways to convey messages instantly to the website visitors. In addition, with various editing and configuration tools available just as in every block along with the ones mentioned before, the alert messages can be customized on its content, appearance, and impact on the reader. In the next section, we will be discussing the rating block under the inner content blocks of the website builder.

Rating block

Rating is a method used by websites to show how satisfied customers who have purchased products or services are. Furthermore, on a website, companies can put forward certain certifications provided on the product or service from a regulation board or an authority of concern. Upon choosing the **Rating** icon from the web page editing menu, you can drag and drop it in the web page, as shown in the following screenshot:

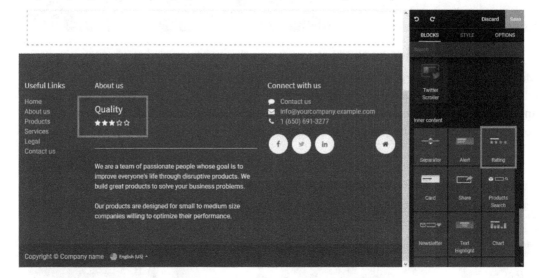

Figure 6.5 – Rating block icon in the web page editing window

Initially, default content or setting will be shown, which can be edited and modified as per your requirements. Moreover, you can change the Icon shape available, modify the color of the active and inactive ones, change the score and modify the size of the block. The following screenshot depicts the editing menu highlighted where these options are available:

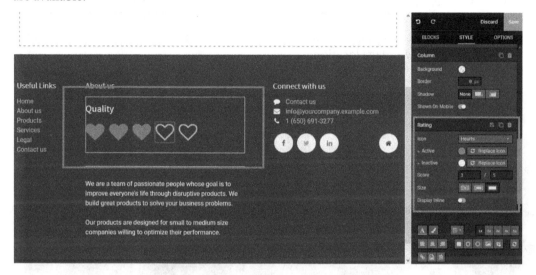

Figure 6.6 – Web page after alert block editing

The rating block is one of the best usable tools in website building as it can be used to showcase a certain quality or unique achievement in a quantitative way. For example, if some governmental authority or a popular magazine has voted you as one of the best companies based on ratings by the public, you can mention it on the website using this tool. Furthermore, there are various styling and configuration options available in the block just as in all other block operations of the Odoo website builder. Now let's move on to the next section, where the card block tool under the Odoo website builder will be explained.

Card block

The **Card** block under the inner content block is a useful tool in website design as it provides you with a designated area in the web page to define the contents in a brief format. Moreover, the card block can contain designated information on the company or the products or services. You can access the **Card** block icon from the web page editing menu, as shown in the following screenshot. In addition, after choosing the **Card** block, just as with any other block operation tools in the Odoo website builder, you can drag and drop it to the desired web page location:

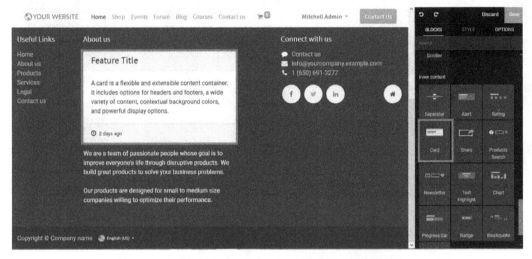

Figure 6.7 – Card block icon in the web page editing window

Once the Card block is operative in the web page, you will initially be displayed with certain default content and block size, which can be modified using the column options available, as shown in the following screenshot. Furthermore, you can change the content, **Style** of the block, **Background**, **Color** configuration, **Alignment**, and many more aspects to suit as per the website design required:

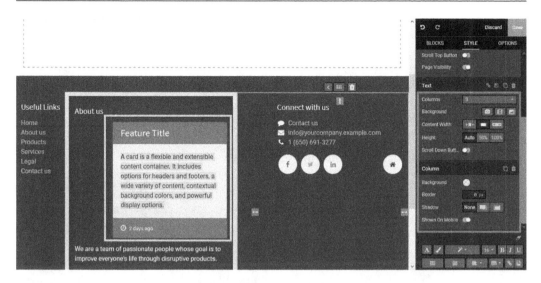

Figure 6.8 – Web page after card block editing

With the use of the Card block, you can specify and describe the contents in the web page. Moreover, it can be used as a medium to describe the various promotional as well as important aspects of a product or a service, along with a mode to convey information to the web page visitors. Additionally, there are various configuration as well as style editing options available as in all other block operations of the Odoo website builder to make the Card block more illustrative and attractive to the visitor. In the next section, we will be discussing the share block tool under the inner content block of the Odoo website builder.

Share block

The influence of social media has revolutionized and modernized the way we view content and exchange information. Although social media platforms concentrate on personal use, they can be the best marketing and promotional tool for a company. Moreover, you could list the company, create a page, and post news on social media platforms. The Odoo website builder **Share** block will allow you to have social media icon buttons on the web page. You can simply click on the block tool and drag and drop it to the desired location on the web page, as shown in the following screenshot:

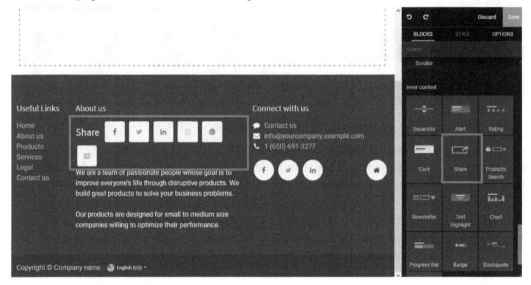

Figure 6.9 – Share block icon in the web page editing window

Initially, you will be provided with a default style of description and the icons will appear as shown in the previous screenshot. However, it can be modified and edited as per your requirements with the configuration options available as shown in the following screenshot. Here, you can set up the **Alignment**, **Title Position**, **Layout**, **Size**, and **Color** of the block. Moreover, there are various other options as well as style configuration tools available just as in the other block tools of the Odoo website builder:

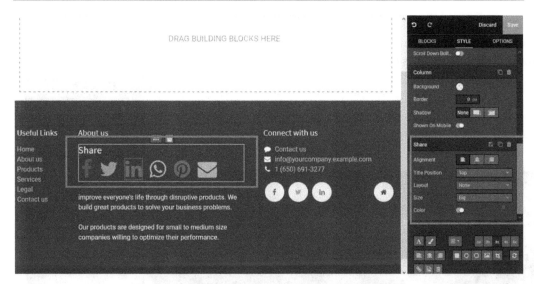

Figure 6.10 – Web page after share block editing

Furthermore, you can configure the social media pages of the website in the backend of the platform as done while creating a website. After configuring the various social media platforms in the web page, the visitor visits the website and they can click on these icons, directing them to the social media page of the company. For example, if the visitor wants to view the Facebook page of the company rather than opening it from social media, they can go directly to the website and click on the icon. As of now, you will have an understanding of the share block tool of the Odoo website builder so now let's move on to the next block under the inner content – the product search – in the next section.

Product Search block

A company should be providing content on their web page based on the products and services they provide to the customer. Moreover, these should be listed on a web page for the customers to select and order them online or purchase them from the retail location. Therefore, the need for a search menu is vital in the website if it has multiple products and a long list of services to showcase.

The Odoo website builder has a **Product Search** block tool for you, which can be dragged and dropped in the desired web page as depicted in the following screenshot, usually in the products or services menu:

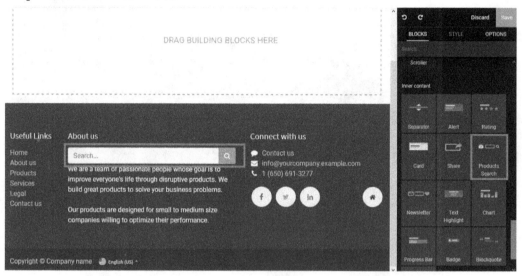

Figure 6.11 – Product search block icon in the web page editing window

Moreover, the search results menu can be configured as to how it should be displayed, such as the listing of the products in a particular order, **Suggestions** for products, a **Description** of the products, **Price**, and **Image**, with the configurable options available in the editing menu, as depicted in the following screenshot:

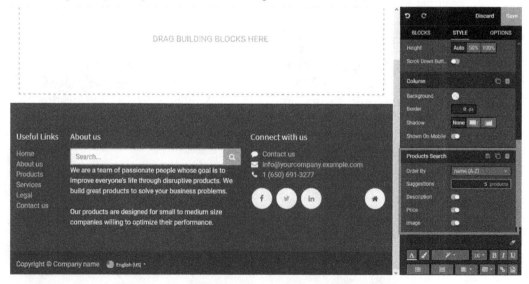

Figure 6.12 – Web page after Product Search block editing

Furthermore, the product search block is one of the extremely useful tools available in the website. With the use of this block, you can provide provision for the visitors to search for their desired products from a long list of products as well as services available from your company. Additionally, there are various style configuration and editing options available, through which the product search block can be made recognizable and as per the design of your web page. Now let's move on to the next block in the inner content block – the newsletter block – in the next section.

Newsletter block

Newsletters are one of the vital aspects of a company, as they serve as a medium to advertise new products, services, the company agenda, and other in-house operations to the customers. Today, the use of hard copy newsletters is being reduced and the online newsletter methodology has been booming. The Odoo website builder recognizes the vital need for the **Newsletter** tool and has allocated a dedicated block to describe and configure the subscription option in the web page. You can select the Newsletter block from the Odoo website editing menu and drag it to the desired location on the web page and place it, as depicted in the following screenshot:

Figure 6.13 – Newsletter block icon in the web page editing window

Furthermore, the **Subscribe** button available for the newsletter subscription should be configured, which can be done by double-clicking on it. Upon selecting to configure, you will be presented with the following window, where the **URL or Email** can be configured. In addition, the button **Type**, **Size**, and **Style** can also be set up:

Link to   ×

| URL or Email | # | Preview |

Hint: Type '/' to search an existing page and '#' to link to an anchor.

Subscribe

Type   Link Primary ✔   Secondary

Size   Medium   ⌄

Style   Default   ⌄

⊂⊃ Open in new window

Save   Discard

Figure 6.14 – Smart icon configuration window under the Newsletter block

Moreover, in a real-time company operation, you will have multiple newsletters in operation, therefore each one should be configured in the respective web page. This can be done by selecting the block and choosing the editing window as depicted in the following screenshot and choosing the option to **Change Newsletter**:

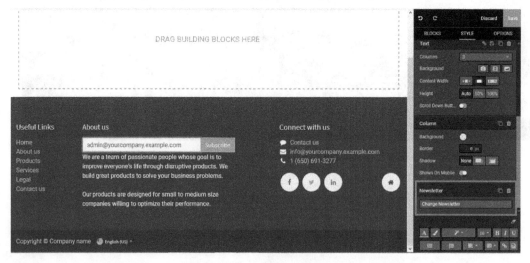

Figure 6.15 – Web page after product Newsletter block editing

Upon selecting to change the newsletter, you will be presented with a subscription menu, where the respective newsletter can be selected from the list as depicted in the following screenshot:

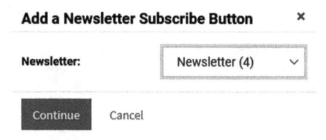

Figure 6.16 – Pop-up window to choose the newsletter

In addition, the newsletters for your company can be published on a weekly, monthly, quarterly, half-year or yearly basis, which mainly depends on the decisions of your company. Whatever frequency you choose the newsletter block of the Odoo website builder will support the operation, allowing website visitors to subscribe to it using this block. Furthermore, there are various other configuration as well as style editing options available, enabling you to configure the block to suit the required appearance of your design and company website. In the next section, we will be discussing the next block type under the inner content block – the text highlight block.

Text highlight block

The website is the best information provider to the customer. Moreover, it's the best way to pass information about the company to visitors to the web page. Therefore, it's vital to mark the important contents, which should be highlighted to grab the attention of the reader. The Odoo website builder has a **Text Highlight** block to provide highlightable content to the web page.

You can choose the Text Highlight block from the web page editing window as depicted in the following screenshot and drag and drop it to the desired location on the web page:

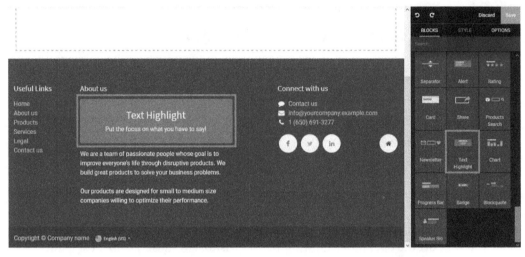

Figure 6.17 – Text Highlight block icon in the web page editing window

Initially, default content will be provided by Odoo for the block to give you an understanding of the design aspects. Additionally, it can be edited and configured further using the options available in the existing window such as the **Alignment**, **Width**, and **Background**, as depicted in the following screenshot:

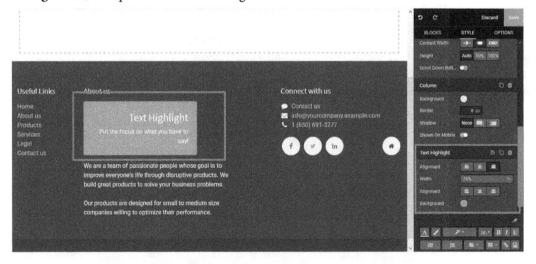

Figure 6.18 – Web page after product Text Highlight block editing

Highlighting text is one way to provide an alert message to web page visitors. In addition, it can be used to provide certain official announcements as well as highlightable contents regarding the operations of your company in the website. The Odoo website builder text highlight block will provide you with various other various style and configuration options available just as in all other website builder blocks of Odoo. In the next section, we will look at the chart block of the Odoo website builder.

Chart block

Charts are tools that can be used to describe the quantitative analysis and data of the company-related aspects to the website visitor. Moreover, a chart on the web page will allow you to provide a comparison of your company with competitors in similar markets. The Odoo website builder provides you with a **Chart** block tool that can be easily dragged and dropped into the desired web page location, as depicted in the following screenshot:

Figure 6.19 – Chart block icon in the web page editing window

Upon choosing the Chart block, it is shown with a default chart to provide you with an insight into the operations of the block. Furthermore, these details can be edited and configured based on your requirements. Moreover, you have options to provide a **Background** to the chart, choose the **Type** and **Alignment**, enable the **Tooltip**, and configure the number of parameters in operation, along with their value. In addition, there are various data editing tools such as **Data Color**, **Data Border**, and **Border Width**, as depicted in the following screenshot:

Figure 6.20 – Web page after chart block editing

Quantitative analysis on product sales, growth, and progression of the company are examples of good charts that can be shown on the company website using the Chart block tool. In addition, certain worldwide or region-based statistics concerning the company or the sector of company operation can also be added to make the visitor understand the trends of the past and the ongoing ones. In addition, you can modify the appearance of the block with the various style and configuration options available just as in all other block tools. Let's now move on to discuss the progress bar block under the inner block content label in the next section.

Progress Bar block

Indicating the progress of the company on a website is the best way to communicate the status of operations. The Odoo website builder has a **Progress Bar** block that can be used to indicate the process of an operation with ease. You can choose the Progress Bar block from the web page editing menu and drag it to the desired location as depicted in the following screenshot:

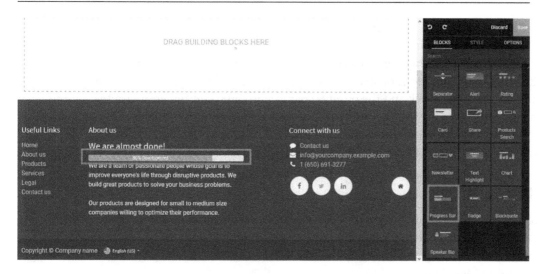

Figure 6.21 – Progress Bar block icon in the web page editing window

Moreover, the contents of the block can be edited by using the various configuration options available in the web page editing menu such as the **Value** of the progress bar, **Display** types, and **Color**, and it can either be **Striped** or not along with the option to enable and disable **Animation**:

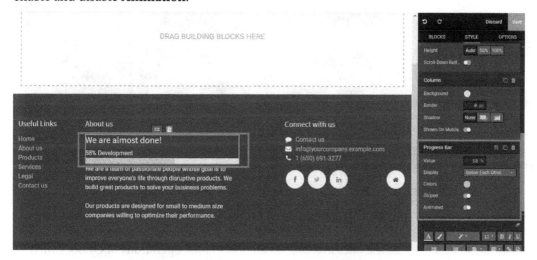

Figure 6.22 – Web page after Progress Bar block editing

The Progress Bar can be an ample support and status depiction tool of the certain aspects of company functioning to the web page visitor. For example, it can depict the function of the status of the transitions of a company to the newer setup or it can be used to depict the processing operations of an aspect that is ongoing. Additionally, you can change the appearance of the block and the progress bar as per your requirements with the options mentioned before or with the configuration and styling options available. Let's now move on to the next block tool under the inner content block classification – the badge block – in the next section.

Badge block

You can provide certain badges of achievement or certification on the company website using the **Badge** block tool on the Odoo website builder. Moreover, you can choose the **Badge** icon and drag and drop it in the desired location on the web page. See the following screenshot for a demonstration:

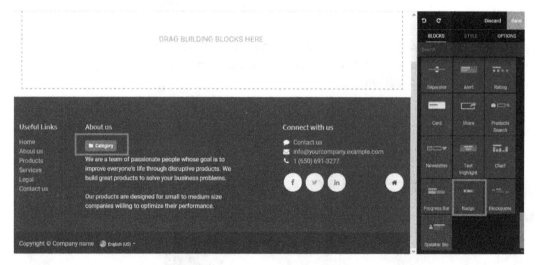

Figure 6.23 – Badge block icon in the web page editing window

Additionally, you can configure the badge based on the **Color** requirement and the content that it should be displaying. Furthermore, there are various other styling and configuration options, like the ones available in all other block operations of Odoo, which can be used to configure and design the web page as per your requirements:

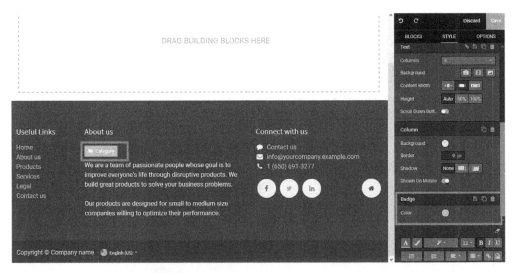

Figure 6.24 – Web page after badge block editing

As of now, you will be clear about the badge block of the website builder so let's now move on to the next section, where the blockquote block will be examined.

Blockquote block

Quotes are an eminent methodology that can be used to express the motive and the methodology of the company. On your website, you can provide quotes from eminent personalities or your employees, previous employees, and customers. Odoo has a **Blockquote** block dedicated to defining the quotes on the web page, which can be obtained from the editing menu, as depicted in the following screenshot:

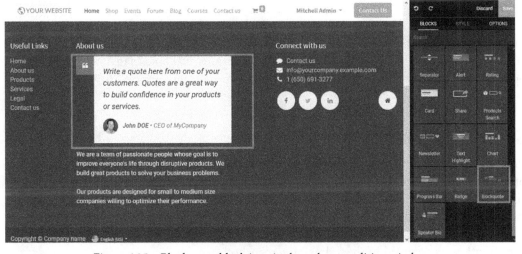

Figure 6.25 – Blockquote block icon in the web page editing window

In addition, upon selecting the Blockquote block, you will be initially provided with default content to provide you with an understanding of the same. Moreover, you can edit and configure this block to suit the operational need of your web page and your content requirement. Furthermore, there are editing options available to assign the **Width**, **Alignment**, **Display**, **Background**, **Screenshot**, **Position**, **Parallax**, and **Colors Filter** options to configure the block, as depicted in the following screenshot:

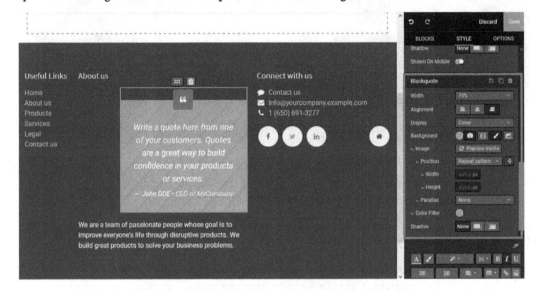

Figure 6.26 – Web page after blockquote block editing

Describing the various attractive quotes and contents concerning the company operation or the aligned sector will be informative to the web page visitors. A dialog quote from an employee, as well as a customer appreciating the company along with its products and services, will provide a better impression of your firm to the customer. Moreover, you can make the Blockquote block more attractive by using the various options mentioned before and with the various other style and configuration options available as in all other block operations. In the next section, we will describe the speaker bio block of the Odoo website builder.

Speaker Bio block

Providing a speaker bio or employee bio on the website is vital for a company, as it can provide certain background information about the employees and the person who is about to speak at a particular event organized by the company. The Odoo website builder has a **Speaker Bio** block that can be used to provide a column in which the speaker bio can be defined. Upon choosing the block, you should drag and drop it in the desired web page location, as depicted in the following screenshot:

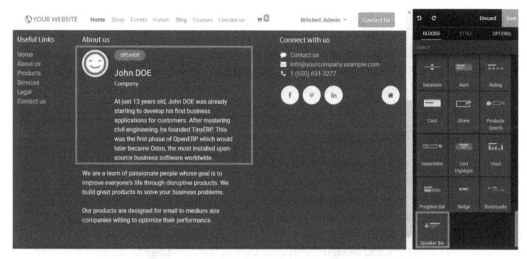

Figure 6.27 – Speaker Bio block icon in the web page editing window

Furthermore, you can provide an image of the speaker, as shown in the following screenshot, and the contents, along with the speaker's name, can be changed from the default one provided initially to give an overview of the operations:

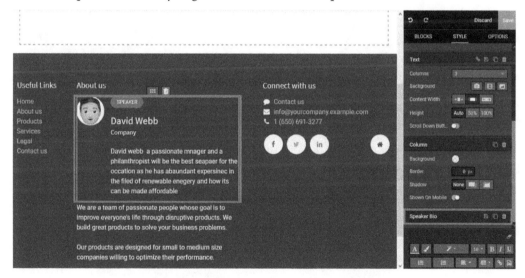

Figure 6.28 – Web page after Speaker Bio block editing

Moreover, there are various types and configuration options available just as in all other blocks of the Odoo website builder.

The Inner Content block will provide you with ample tools that can be used to define the contents of the web page. Moreover, these blocks will help you to highlight the specific and important contents that should be described on the website to catch the attention of the visitors. In addition, these blocks will act as a platform to define the contents of the web page in an orderly and attractive fashion.

Summary

In this chapter, we gained an understanding of the Inner Content block of the Odoo website builder and the types allocated under the classification. Moreover, we had an insight into the design aspects of the web page contents using the Inner Content blocks to make the web page more attractive and better arranged. Furthermore, the Inner Content blocks will allow you to define the supporting aspects of the company in the website, with the ability to modify and design the appearance of the web page to make it the most suitable one for the company.

In the next chapter, we are going to learn about HTML/CSS/JavaScript Editor used in the Odoo website builder.

Questions

1. How many types of Inner Content blocks are available?
2. Which block can be used to highlight content in the web page?
3. Which blocks can you use to describe the employee or customer feedback on the website?

Further reading

- *Working with Odoo* by Greg Moss, Packt Publishing
- *Learn Odoo* by Greg Moss, Packt Publishing

7
HTML/CSS/JS Editor

In the previous chapters, we discussed the various methods of website building using the Odoo website builder. Moreover, we specified insights on blocks and their usage in website building operations, including style and options configuration, which can be customized to suit your website and company preferences for the appearance of the website. In addition, we discussed various advanced and specific tools used in website building using Odoo's website builder.

In this chapter, we move onto the technical aspects and the programming side of website building, which will be described from the following aspects:

- Introducing HTML/CSS/JS Editor
- Using HTML/CSS/JS Editor

By the end of this chapter, you will be well equipped with the technical knowledge to program the various aspects of a website using the website builder in Odoo.

Technical requirements

Install the Odoo platform on a system with the capability of processing the operation at a moderate and acceptable speed. In addition, a basic knowledge of programming as well as coding the various aspects of a website will be beneficial. Moreover, a deep knowledge of the various aspects of website building using the Odoo website builder is essential.

Introducing HTML/CSS/JS Editor

HTML, CSS, and JS are the programming languages that are used to run the web. Almost all applications and services available today utilize these forms of programming languages to perform their operations. These are user-friendly applications that can be learned and understood with ease with a basic knowledge of coding. To people new to the field, it may feel a bit difficult in the beginning, but gradually, it will be easier to pick up.

HyperText Markup Language (**HTML**) is one of the most commonly used languages and the main programming language that helps you to design and display documents and content on a web page or a website. HTML code is always supported by CSS and by the scripting language JS. **CSS** stands for **Cascading Style Sheets**, which is used for configuring the presentation of a document on a web page in HTML. Moreover, **JS** stands for **JavaScript**, which is a programming language used for web development purposes and is used to add dynamic as well as interactive content and elements to HTML.

As these are some of the most used programming elements to design, develop, and deploy a web page, the Odoo platform recognizes its vitalness and has included these elements in the website editing tool. Moreover, the Odoo website builder application has web page configuration operations that can be enabled to perform operations in HTML, CSS, and JS content using the available editor tool. Furthermore, the tool has the combined capability to perform all the operations of these programming aspects and can be accessed from the frontend of the platform and directly from a web page itself.

The operations of programming and configuring a web page using the Odoo website builder are carried out by a collective effort of all three elements: HTML, CSS, and JS. Moreover, programming a web page for the company website will include web development with the help of these three collective elements functioning together with the ultimate goal of developing an attractive company website for you. The following block diagram is a depiction of the collective effort operation for the web development, resulting in a website:

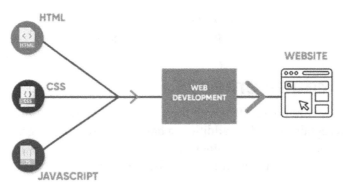

Figure 7.1 – Block diagram of web page development

The web page design and building of the website using the Odoo website builder can be defined as chronological operations that follow certain principles in functioning. Moreover, with the use of HTML/CSS/JS Editor tools, a website developed using the functional aspects can be modified and restructured technically rather than using the block and web page design tools that were discussed in the previous chapters of the book. In web page editing using the HTML/CSS/JS Editor, you will initially draft a design for the web page and design and build one. If your website consists of more than one web page, all are built individually or in collaboration. Using all these web pages, the company website is set up and running. In the case of further edits and modifications, these can again be done using the HTML/CSS/JS Editor for the respective web page and the modifications can be saved and put on to the website.

The following figure shows a flow chart that showcases the workflow of the operations as discussed:

HTML, CSS & JAVA   WEB PAGE   WEBSITE

Figure 7.2 – Flow chart of website building

In the following sections, we'll understand HTML, CSS, and JS programming aspects in brief.

HTML

HyperText Markup Language is one of the standardized languages for the creation and design of web pages. The language is responsible for providing a structure for a web page. Therefore, there are various elements, constraints, and attributes in operations to consider while designing a web page using it. Moreover, it provides information on how the elements and contents of a web page are to be displayed in a browser.

Let me give you an example of HTML file content that will allow us to have a clearer understanding of the topic:

```
<!DOCTYPE html>
<html>
<head>
<title>Page Title</title>
```

```
</head>
<body>
<h1>My First Heading</h1>
<p>My first paragraph.</p>
</body>
</html>
```

Let's look at the example and its code in detail:

- `<!DOCTYPE html>`: Indicates that the document is in HTML5 format.
- `<html>`: Acts as the root element of every HTML file.
- `<head>`: Contains information about the web page.
- `<title>Page Title</title>`: Contains the page title, which is indicated at the head of the web page.
- `<body>`: All the elements and the contents along with the various configuration options are defined here.
- `</h1>`: Defines the heading.
- `</p>`: Describes the contents in paragraphs.

An HTML element is something that is contained in every code of operation. It will contain a start tag name, content, and an end tag name arranged in the same order in the code. The elements in an HTML file can be well explained with the help of an example such as this:

```
<h1>My First Heading</h1>
<p>My first paragraph.</p>
<br>
```

Let's discuss the preceding code line by line. The `<h1>` in the first line indicates the start tag of the code and the following My First Heading is the heading for the content that you have to provide, which is the element of the operation. Finally, `</h1>` indicates the end tag for the line and the process has been ended.

In the second line, the `<p>` indicates the start tag specified for the paragraph design, and the content is provided just as indicated as My first paragraph, which is the element of the function where the entire paragraph content for the web page is provided. Additionally, `</p>` is the end tag of the paragraph description.

In the final line, the code is indicated as `<br>`, which demonstrates breaking the operation line. However, as it is the break operation line, there is no element or end tag line available. Moreover, the element has the sole purpose of creating a line break in the block content being described.

Web browsers such as Google Chrome, Mozilla Firefox, Internet Explorer, and many more are used to read the HTML files present in a web page or a website and produce the apt document or display the views as intended by the web page developers. Moreover, browsers use elements in the code just as described in the previous example and display it based on the commands provided in the tag section of the code. Additionally, each content description, the mode of the display, as well as the style and configuration will be specified in the HTML code, making the browser options exactly follow it. The editor also has various web layout options that will bring in a website structure and are used to display the information and the contents described are suitable to the viewer as well as the website owner.

Under the web layout, there will be a description of the various web page editing tools available, which are described and depicted on the website. Among these, the frontend layout described will help you to describe the graphical aspects of the web page. Additionally, the main frontend layout options available will help you to provide a design for the main aspects of the web page. The footer language selector configuration will provide website visitors with a language selector option in the footer of the website.

The footer language selector flag can be a useful tool on a website for visitors as they will be able to select the website language using the flag options that have been described. Furthermore, you can describe the main layout of web pages in the XML editor, which can be configured based on the contents and the design elements that have been described. Additionally, there are various other configuration tools and options that can be configured in the XML editor menu as per your requirements. As of now, you will have a generalized understanding of HTML files and how coding operations are conducted, therefore, in the next section, let's move on to CSS language operations. Now, you will have a generalized understanding of HTML files and how coding operations are conducted, therefore, in the next section, let's move on to CSS language operations.

CSS

CSS provides an insight into HTML elements and how they should be described on a web page. Although HTML is considered the base programming language for a web page as well as website development, it is limited in its design as well as styling capabilities. Later, from the development of the HTML methodology, the CSS language was introduced into the HTML, acting as the savior and the problem solver for the styling and design needs of web pages.

CSS is one of the best solutions for web page design as it allows design and management capabilities on multiple web pages all at once. Moreover, CSS files will store external file sheets that will act as the design element in HTML files as well as web pages. In addition, the styling elements of CSS files in HTML files are responsible for web page design, layout, variation in the display, alignment, and configuration for various devices and screen sizes, along with various other appearance configuration options.

HTML was introduced and invented to display the contents of a web page and not to meddle with the formatting and appearance aspects of a web page. However, in one of the versions of HTML (3.2 to be exact), they introduced the editing aspect to the HTML file, which was later ruled out as one of the catastrophic developments in the system due to the complexities it brought into operations. Therefore, a new aspect of CSS was introduced to solve the issue. Moreover, CSS removed the configuration and styling aspects from HTML content, again providing the same efficiency in operations, just concentrating on loading the contents to the web page.

```
H1{ color:green; font size:11px; }
H1: Selector section
```

The preceding code indicates the heading or the HTML element that you want to configure.

```
{ color:green; font size:11px;}: Acts as the declaration
content
```

The declaration block will have one or more declarations injected into the code, indicated in brackets as shown in the preceding code. Each of the declarations will include the property name similar to the HTML file and the value or the entity it has been changed to. Additionally, multiple declarations can be defined using semi-columns as per your need under the same selector.

The operation of CSS files can be well explained with the help of an example such as the one mentioned next:

```
p {
   color: red;
   text-align: center;
}
```

Let's have a look into the example and how the code is used in detail:

- P: Indicates the selector in the CSS file and points out the location in the HTML file or the content.
- color: The property or the aspects of styling that need to be modified to suit the color coordination required.
- red: The entity to which the property value is to be modified.
- text-align: Another property in operation in the counts. It also adjusts the alignment.
- center: The entity to which the alignment of the content is changed.

As of now, you will have an understanding of CSS files and their formatting, as well as configuration aspects. In the next section, let's move on to understand JS files used in HTML operations.

JS

JS stands for JavaScript. It's considered as one of the most powerful and most influential programming languages. Moreover, it has been used as the programming language for the web and is considered one of the vital programming languages that all developers should have an understanding of. The use of JS with the integration of HTML files will provide you with advanced capabilities in web page design and development aspects.

In an HTML operation, HTML files are used to describe the contents of the website, CSS files provide layout and other appearance aspects of the HTML content, and JS is used to describe the behavioral aspects of the web page and the entire website in operations. JavaScript is a free application tool and is already running in your browser, system, and even in your smart gadgets and devices.

Let's understand the operation of JS in an HTML file with the help of an example:

```
document.getElementById("demo").innerHTML = "Hello JavaScript";
```

getElementById is one of the most commonly used and is used to find an HTML element using the ID. So, in the preceding example, the command will search for HTML content with the ID demo. The inner HTML content is modified into "Hello JavaScript".

You can change the style of an HTML element using JavaScript by editing and configuring the CSS file. This can be explained with an example:

```
document.getElementById("demo").style.fontSize = "35px";
```

Using the preceding code, JS will search for the "demo" file and change the font size to 35px. Various other style configuration changes and editing can be performed and modified based on the requirement using similar command operations. In the next section, let's understand the various aspects of using the HTML/CSS/JS Editor in Odoo website building operations.

Using HTML/CSS/JS Editor

The Odoo website builder has an HTML/CSS/JS Editor tool available at the frontend of the platform, accessible upon visiting the web page. This editor acts as a real-time modification tool where web pages, websites, and all aligned aspects described on a web page can be edited, configured, and modified. Moreover, with a basic understanding of HTML, CSS, and JS, you can configure and edit the various aspects.

To access the HTML/CSS/JS Editor, go to the website of the company from the backend of the platform and you will be shown the home page of the website, which can be configured. From the home page or the respective web page where the editing is to be done, select the **Customize** option from the dashboard as shown in the following figure and you will be able to choose **HTML/CSS/JS Editor** from it:

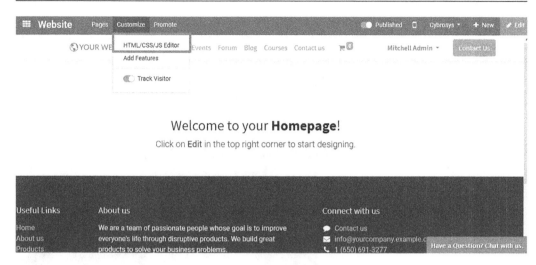

Figure 7.3 – HTML/CSS/JS Editor option on the website

Upon choosing **HTML/CSS/JS Editor** on the screen, you will be taken to a code provider window as shown in *Figure 7.4*. Here, you can modify the existing lines of code as well as inserting new ones and removing the ones not required.

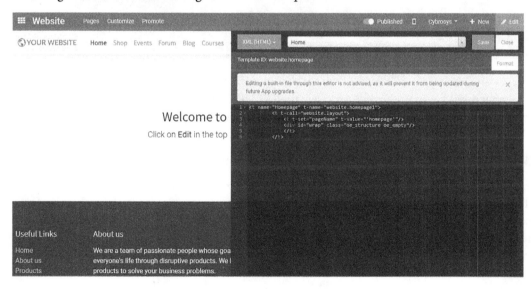

Figure 7.4 – The HTML/CSS/JS Editor window

In addition, the **HTML/CSS/JS Editor** window has considerable configurational and operating tools. You can choose either of these options: for HTML aspects of the web page, choose the existing tools on the XML file, or the JavaScript elements of the described constants, or choose the SCSS (CSS). Now, in the previous session, we described the usage aspects of each of the editing files and languages. The operations here are conducted similarly:

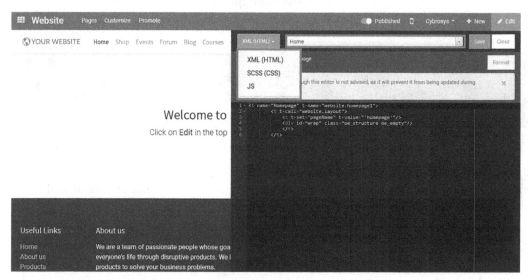

Figure 7.5 – Menu options in HTML/CSS/JS Editor

Let's now move on to understanding how the HTML/CSS/JS Editor can be used well to configure the web page content of your website.

Setting product name tag alignment

In the Odoo website builder using the HTML/CSS/JS Editor, you can modify the alignment of the product description displayed on the e-commerce page of a web page, for example, the shop page. This feature is best used as a customization tool to align product descriptions on web pages too. You can similarly configure the alignment of the content of the web page.

The CSS editor can be used for this configuration. Additionally, the following code will provide you with the option to configure the product description alignment:

```
.oe_product_cart .o_wsale_product_information{
    text-align: center !important;
```

This code can be copied to the SCSS (CSS) editor tool as shown next:

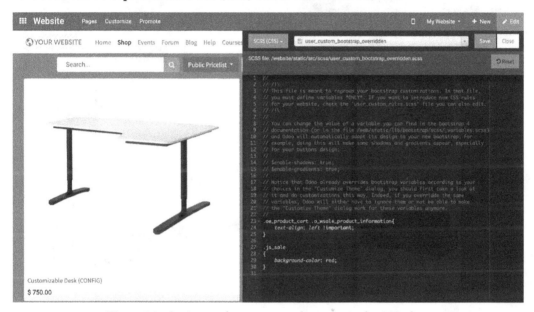

Figure 7.6 – Setting product name tag alignment in the CSS editor

Initially, by default, the text alignment of the product information will be center-aligned, which can be seen in the preceding code. `o_wsale_product_information` is used to drive the attention of the command to the product description on the shop page of the website. Additionally, the text alignment is centered so that the contents are centered just as showcased in the following screenshot. The `important` class is used for the operation to be inevitable than any other non-important class in the code.

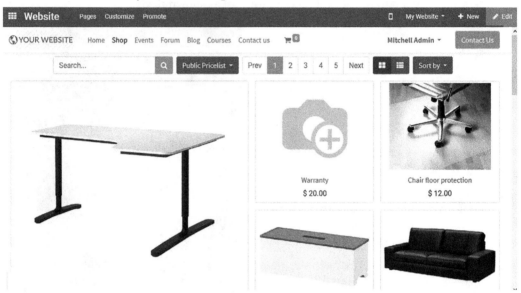

Figure 7.7 – The product description is center aligned

Now, to right-align the product description constant, the same code used for the center-alignment configuration is used. However, the text-align option is provided to right and the rest of the code remains the same, just as described in the following:

```
{
.oe_product_cart .o_wsale_product_information{
    text-align: right !important;
}
```

The right alignment of the product description can be described as one of the normalized style elements and will be based on the descriptive and design aspects of the web page according to you. The following figure depicts the style change in the product description on the shop's web page according to right alignment:

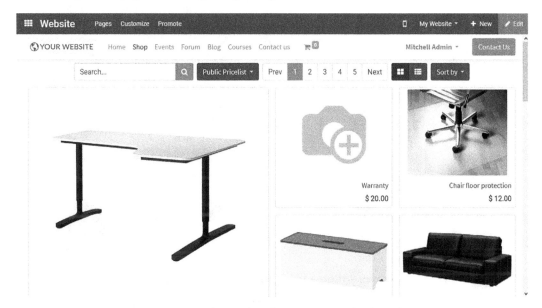

Figure 7.8 – The product description is right-aligned

Now let's move on to understanding the left-alignment aspects of the product description, as it's one of the most common aspects of an e-commerce web page since it provides more space to describe the contents compared to others and a definite standardized state accepted all around the world. To describe the alignment as left alignment, you can use the same code as that of the center and the right alignment, as discussed earlier. However, we make a slight change to the text-align description, to left, as in the following command:

```
{
.oe_product_cart .o_wsale_product_information{
    text-align: left !important;
}
```

The following screenshot showcases the resultant web page after the left-alignment operation is conducted on the shop page for the product description.

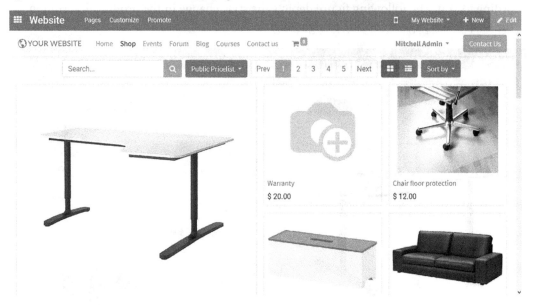

Figure 7.9 – The product description is left-aligned

As of now, you will be clear on how alignment operations for product descriptions are conducted. Similar operations and code can be used with further modifications to deal with the alignment operations of the other web page content on the website. In the next sections, let's move on to understanding how to provide a background color for a web page.

Background colors for web pages

Providing a background color for the web page is an efficient way to design a website to be more attractive. Today, numerous websites are functioning, and to be distinctive from them, content and appearance can play a crucial role. The Odoo website builder provides various provisions using block operations as mentioned in the previous chapters. However, if you need further generalized and common editing of the web page constant, the HTML/CSS/JS Editor will come in handy.

The addition of background color to a web page can be easily done with the help of a line of command providing the adequate configuration to assign web page backgrounds. The following code corresponds to the background change operations of the web page:

```
.js_sale
{
    background-color: red;
}
```

In the code, the background color for the page is depicted as red and you can provide the color name that you wish to have for the web page of the website. Moreover, images, as well as background illustrations, can also be defined. Anyway, the background color will be applied only to the outer spaces, as shown in the following screenshot, where the product blocks are isolated and no color is provided:

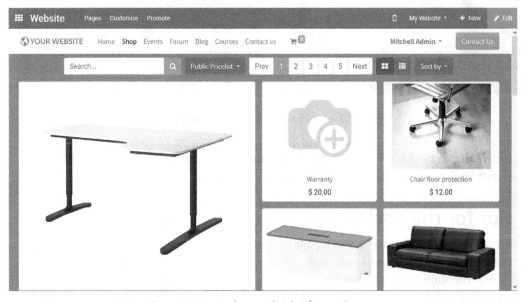

Figure 7.10 – Background color for a web page

In addition, the implications can be seen while selecting a product, as the background or the entire shop window is set as red. The background change is made only to the web page and the contents in it are isolated as in the following figure:

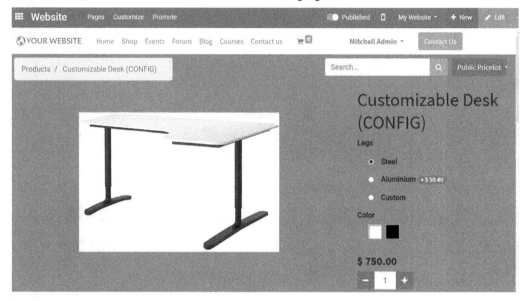

Figure 7.11 – Background color for product page

The background color configuration will act as a medium to make your website more attractive. It can be used along with the background on the product block, as described in the following sections, to create an illustrative and attractive web page.

Color for the background and product block

Color for the background of the web page as well as the product description provides an attractive medium to design a web page. Moreover, you can use a company's standardized color coordination to match it with the website's color. It will be a marketing tactic as well as a branding parameter. The operation can be configured similarly to as described in the *Background colors for web pages* section. Let's now have a look at the code used for configuring it via the HTML/CSS/JS Editor:

```
.js_sale{
    background-color: red;


}
```

```
.oe_product_cart .oe_product_image {
    background-color: blue !important;
}
```

In the code, we have two sections, where the initial one is the same as was used in the *Background colors for web pages* section and will provide a background for the web page. In the second part, the background for the product description is provided and is marked with an important class, to showcase that the function is inevitable in the operations. The resultant web page after the background configuration is depicted in the following screenshot:

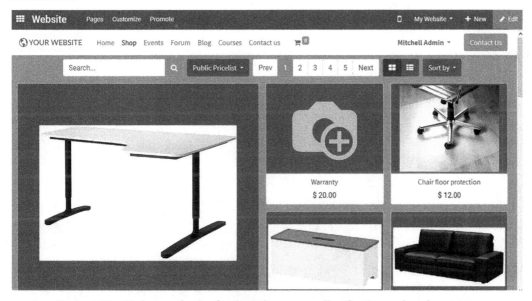

Figure 7.12 – Background color for the web page as well as for the product description

The configuration of background colors provides a way to customize and personalize web pages based on the company's or the designer's perspective. Let's now move on to understanding how to create and provide a banner for the web page.

Providing a banner for the web page

Banners on a web page are an important as well as an illustrative way to convey information and grab the attention of visitors. You can provide banners on a website using the Odoo website builder in two ways. The first way is as discussed in the *Banner block* section in *Chapter 3, Introduction to Blocks – Structure Blocks*, and the other is by using the HTML/CSS/JS Editor. Moreover, a developer with a basic level of programming knowledge can easily add a banner to a web page.

Let's now move on to understanding the HTML aspects of web page editing, using the editor to configure a banner:

```html
<div class="hero-image">
    <div class="hero-text">
        <h1>Cybrosys Technologies</h1>
        <p>Best Odoo gold partner</p>
        <button>Contact us</button>
    </div>
</div>
<div id="wrap" class="oe_structure oe_empty"/>
```

In the code, as you can see, there is a class called to describe the text and a class for the image:

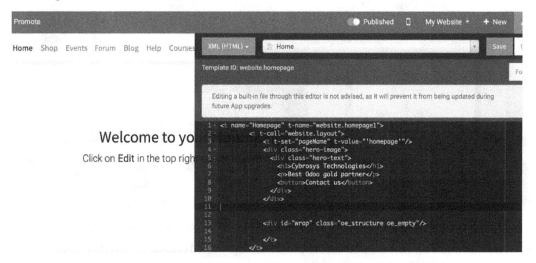

Figure 7.13 – Command in the HTML editor to provide a banner on the web page

Additionally, a heading and constant are being described along with a button for the visitor to be directed to a different web page. The button can be configured based on your requirements. The resultant banner, as per the mentioned code, is depicted in the following screenshot:

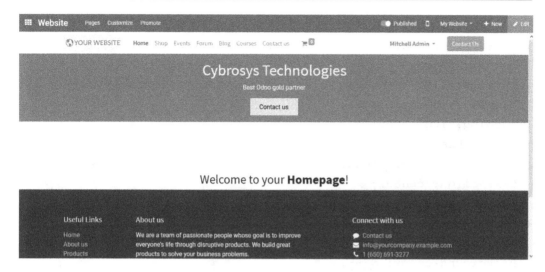

Figure 7.14 – Banner on the web page

Now, the banner will not look good with a background so let's now add an image to it using the CSS editor as its the styling component of the web page and all the styling aspects are configured in the CSS editor of the Odoo website builder. Let's now look at the code for the operation:

```
.hero-image {
  background-image: linear-gradient(rgba(0, 0, 0, 0.5), rgba(0,
    0, 0, 0.5)), url("https://www.chatelaine.com/wp-content/
    uploads/2018/10/Black-Panther-chadwick-boseman-
    e1539272923602-810x608-1539273014.jpg");
  height: 50%;
  background-position: center;
  background-repeat: no-repeat;
  background-size: cover;
  position: relative;
}
.hero-text {
  text-align: center;
  position: absolute;
  top: 50%;
  left: 50%;
  transform: translate(-50%, -50%);
  color: white;
```

```
}
.hero-text button {
  border: none;
  outline: 0;
  display: inline-block;
  padding: 10px 25px;
  color: black;
  background-color: #ddd;
  text-align: center;
  cursor: pointer;
}
.hero-text button:hover {
  background-color: #555;
  color: white;
}
```

As mentioned in the code, the image for the background can be one available in a database or the server or can be added as a direct link to a different web page if the platform has internet access. Moreover, all the other configurations can be described based on the banner as required. The resultant web page after the banner has been added is depicted in the following screenshot:

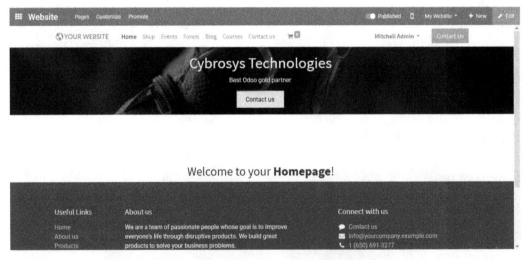

Figure 7.15 – Banner with background image on the web page

In addition, the background for the banner can be a color also, which can be similarly configured, providing the viewer with a delightful web page design, attracting more visitors to the web page.

Contact us page validation

Using the JS editor window of the HTML/CSS/JS editor, you will be able to configure the **Contact us** page as per the validation requirements. Here we are about to configure the description boxes of the **Contact us** page in a way that it will only take as input a number, a word, or special symbols, otherwise an alert will be generated for the visitor. The following code can be used for the **Contact us** page in the JS editor window:

```javascript
// Validate the contact us form fields after each input change
// rather than on form submission
odoo.define('website.contact_us_validation', function (require)
{
    'use strict';
    var reg = new RegExp("^[a-zA-Z ]+$");
    var num_reg = new RegExp("[0-9]")
    var email_reg = new RegExp("^[a-zA-Z0-9+_.-]+@
                               [a-zA-Z0-9.-]+$")
    var Dialog = require('web.Dialog');
    // check whether the name field only contains characters
    $("#contact1").on("change",function(){
        if(this.value){
            if(!reg.test(this.value)){
                Dialog.alert(this, "Name field cannot contain
                            numbers!!..");
                this.value='';
            }
        }
    });
    // check whether the telephone field only contains number
    $("#contact2").on("change",function(){
        if(this.value){
            if(!num_reg.test(this.value)){
                Dialog.alert(this, "Phone Number can only
                            contain numbers!!..");
                this.value='';
```

```
                }
            }
        });
        // check whether the email format is correct
        $("#contact3").on("change",function(){
            if(this.value){
                if(!email_reg.test(this.value)){
                    Dialog.alert(this, "Please check the email
                                format!!..");
                    this.value='';
                }
            }
        });});
```

You can select the HTML/CSS/JS editor from the **Customize** menu and opt for the JS editor window. In the JS editor window, you can provide in the previous code as shown:

Figure 7.16 – Command in the JS editor to configure the search bar

Once the code is added and the changes made to the editor are saved, you can type a number in the **Your Name** field as seen in the following screenshot:

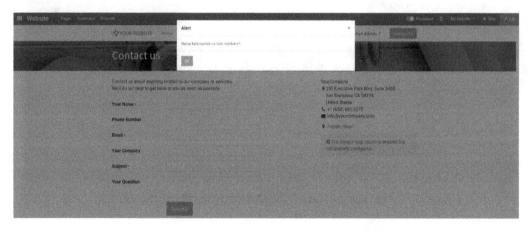

Figure 7.17 – Proving a number in the Your Name field

When you do that, the **Alert** message will pop, and up as you can see in the following image, informs the visitor that **Your name** cannot contain numbers:

Figure 7.18 – Alert message on number being used in the Your Name field

Now try adding alphabets in the **Phone Number** description box as shown next:

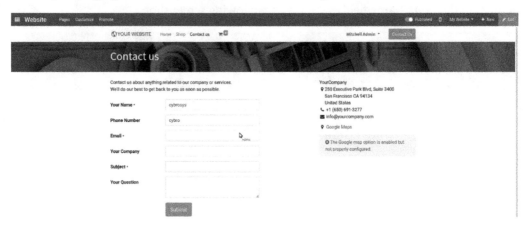

Figure 7.19 – Use of alphabet in the Phone Number field

You will again be prompted with an error message. The following image shows the **Alert** popup window indicating that the phone number can only contain numbers:

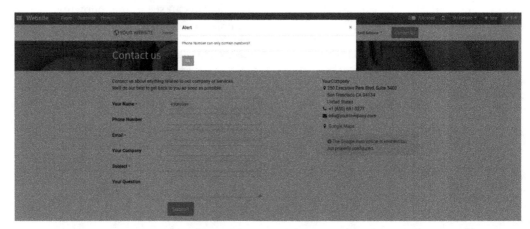

Figure 7.20 – Alert message upon using alphabet in the Phone Number field

Similarly, try adding email address in a wrong format, as shown in the following figure:

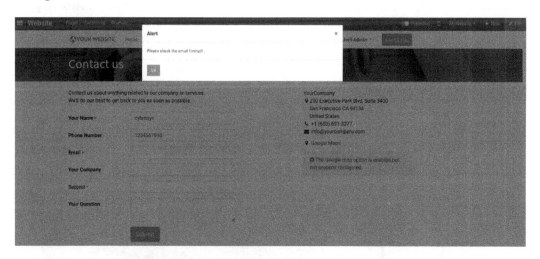

Figure 7.21 – Use of wrong email format

An **Alert** pop-up window will come up, showing that a wrong format of email address has been provided:

Figure 7.22 – Alert message upon usage of wrong email address

Hover zoom on the product block

You might have used various e-commerce websites where the product image depicted gets magnified when we place the cursor over it. The same functionality can be embedded into the e-commerce webpage of your company with the help of the following code:

```
// Zoom product image on mouse hover
odoo.define('website.hover_zoom', function (require) {
    'use strict';    // zoom in the image while mouse hover
    $(".oe_product_image").on("mouseenter", function(){
        $(this).css("transform", "scale(1.2)");
    });    // zoom out the image to original size while leaving
        // the mouse hover
    $(".oe_product_image").on("mouseleave", function(){
        $(this).css("transform", "scale(1)");
    });});
```

You can add this code in the JS editor, which can be obtained by selecting the **Customize Page** option and selecting the **HTML/CSS/JS** editor, and you will be depicted with the editor window. Then select the **JS editor** window and adding the code as depicted in the following image:

Figure 7.23 – Code for hover zoom in the HTML/CSS/JS editor window

Once the code is loaded and saved, when you move your cursor over the respective product, its image will be magnified and projected to you as shown in the following image:

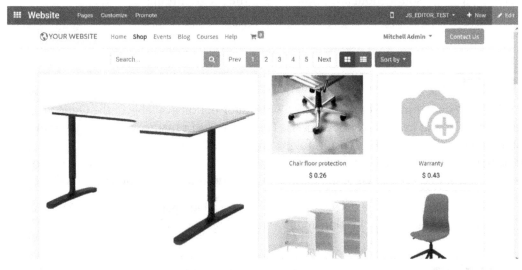

Figure 7.24 – Hover zoom of the product image when the mouse pointer is on the image

Additionally, when the mouse pointer is moved away from the product image, the product will be displayed in the normal size as shown in the following image:

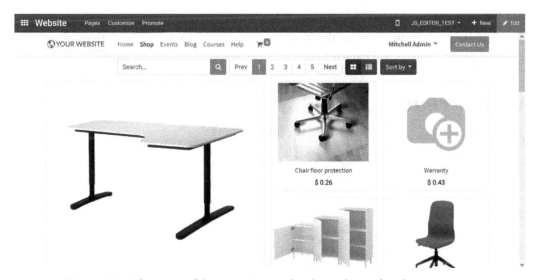

Figure 7.25 – The image of the respective product being depicted without any zoom

There are numerous ways to configure this option in a webpage using the HTML/CSS/JS editor; however, basic programming knowledge is necessary to run these operations. Therefore, for a person without any knowledge in the programming and coding area, it's better to stick with the functional aspects of designing the webpage using the block operations.

Summary

In this chapter, we discussed the HTML/CSS/JS Editor option available in the Odoo website builder and its allocated operations. The tool acts as a further design element in configuring a web page and will be useful in sophisticated design aspects of web pages. As of now, you will have a basic understanding of the HTML/CSS/JS Editor in the Odoo website builder and how it can be configured in operations to design and align a web page.

In the next chapter, we will be discussing how to create a blog page on your website using the Odoo website builder application and how to manage them.

Questions

1. In the HTML/CSS/JS Editor, which menu can be used to configure the various styling elements of web page contents?

2. What do HTML, CSS, and JS stand for?

3. Can we insert an image from the internet or any other social media as the background image of a banner?

Further reading

- *Working with Odoo* by Greg Moss, Packt Publishing
- *Learn Odoo* by Greg Moss, Packt Publishing

Section 3: Practical Tools

In this section, we'll explore other practical tools of the Odoo website builder, including coverage of contact forms, live chat functionality, and more.

This section consists of the following chapters:

8

Creating Your Own Blog Pages

In the previous chapter, we were discussing the HTML/CSS/JS editor, by which we will be able to define and configure the website, its design, and the appearance aspect in the Odoo website builder. Although we have block tools in operation in the Odoo website builder, there will be certain configuration aspects that can be useful in designing the web pages, which can be done using this editor. Moreover, the chapter was focused on educating you on the various aspects of the editor tool and how it can be functional in operation.

In this chapter, you'll learn how to create a blog page on your website using the Odoo website builder. We will mainly focus on the following topics:

- Creating a new blog
- Designing blog content
- Understanding the use and importance of tags
- SEO management

By the end of this chapter, you will be able to create, design, and manage the blogs on your website.

Technical requirements

A basic knowledge of website building using the Odoo website builder is mandatory. People new to the concept will find it harder in the initial stages, but will gradually progress. However, for a person who has read and understood the previous chapter, this will be a piece of cake. In addition, a system with a fully functioning Odoo platform with an acceptable speed of operation will be essential in performing the tasks.

Creating a new blog

Writing and publishing blogs on company websites and external social media platforms is the best way to describe the company's products, their salient features, provide a comparison, and will also be a marketing tool for the company's operations. Nowadays, the presence of blog writers as well as content writers has become an integral aspect of a company's operations as they support them along with **Search Engine Optimization (SEO)** aspects of the websites. Moreover, it can be said that they are the people who showcase the company's products and services in the best and most marketable way possible. Every establishment will have these kinds of support teams for the marketing as well as website description aspects, which will include marketing executives, content writers, website developers, SEO teams, and many more individuals besides, to run the company from the frontend of its operations.

Odoo understands the needs and necessities of having a blog page and allocated tools for the blog operations in the website. Therefore, the platform has a separate blog module allocated to the application menu that can be installed when required. Moreover, the blog module is integrated with all other website-related applications, providing you with a collective approach to management. In addition, the blog module will provide full functionality to the Odoo website builder to design and configure the blogs for the website based on your requirements.

To install the **Blogs** module, select the application module from the home page of the Odoo platform and search for the blogs module. It is also visible under the website category, as depicted in the following screenshot:

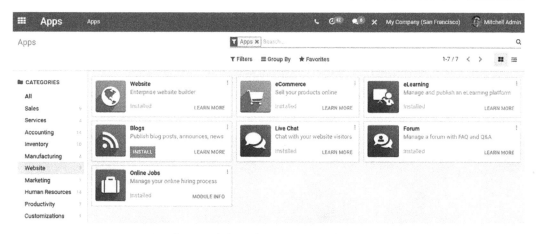

Figure 8.1 – Blogs module in the apps menu of the Odoo platform

You can select the **Install** option available for the module to be installed on the platform. Furthermore, if you need any information regarding the module and want to read about the documentation aspects of it, select the **LEARN MORE** options available. Upon installation, you will be automatically directed to the company's website and the blog page. The blog page, as depicted in *Figure 8.2*, is the result of installing the module from the application menu. Initially, a default constant will be shown to showcase to you how the operation is to be configured if the demo data is enabled upon configuring the platform:

Figure 8.2 – Blog page in the website after installing the application

You can also edit the default contents of the blog page, just as was done in the block operation in the previous chapter. Select the **Edit** option available in the top-right-hand corner of the screen and you will be taken to the web page editing menu. Next, select the content to be edited and this can be modified easily by providing the required content as well as the background image or video available, as depicted in *Figure 8.3*. In addition, as this is a web page, you can configure the tags and describe them in the tag list:

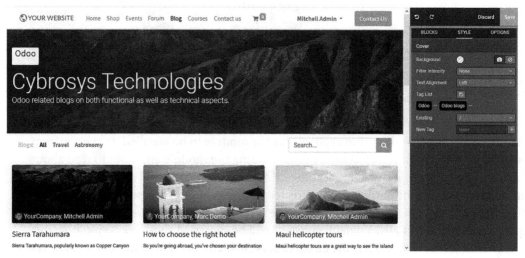

Figure 8.3 – Blog page editing menu

You can configure the blog posting from the backend of the platform by configuring the functional aspects related to it. From the website, the module selects the **Blog post** menu, which can be availed from the + symbol available in the dashboard. Upon selecting the blog menu, you will be presented with all the blog postings as in *Figure 8.4* and as described on the website, along with all the allocated details regarding the **Author**, **Blog** type, **No of views**, **Last contributor**, and the **Last Updated on** date and time:

Title	Author	Blog	Website	No of Views	Last Contributor	Last Updated on	
Beyond The Eye	YourCompany, Mitchell Admin	Astronomy		0	OdooBot	03/08/2021 13:10:36	
Buying A Telescope	YourCompany, Mitchell Admin	Astronomy		0	OdooBot	03/08/2021 13:10:36	
What If They Let You Run The Hubble	YourCompany, Mitchell Admin	Astronomy		0	OdooBot	03/08/2021 13:10:36	
How To Look Up	YourCompany, Mitchell Admin	Astronomy		453	OdooBot	03/08/2021 13:10:36	
How to choose the right hotel	YourCompany, Marc Demo	Travel		467	OdooBot	03/08/2021 13:10:36	
Maui helicopter tours	YourCompany, Mitchell Admin	Travel		246	OdooBot	03/08/2021 13:10:36	
Cybrosys Technologies	YourCompany, Mitchell Admin	Travel		1	Mitchell Admin	03/08/2021 13:15:51	

Figure 8.4 – Blog post menu in the backend of the platform

You can also select an existing blog post for editing it and configuration using the various options available. Additionally, new blog postings can be created by selecting the **Create** option available in the window, which will direct you to the web page, as depicted in *Figure 8.5*. Check out the blog post creation window where you need to provide a **Blog** name, as well as **Title**, **Sub Title**, and **Tags** information allocated along with it, and the website in which the blog should be published if your platform is functioning for multiple company operations.

Additionally, under the publishing options, you can configure the **Author** and **Publishing date** fields. Furthermore, the **Last Contributor** name and the **Last Updated on** date will be available for the blogs already published, and not for the new blog postings that are in the process of creation:

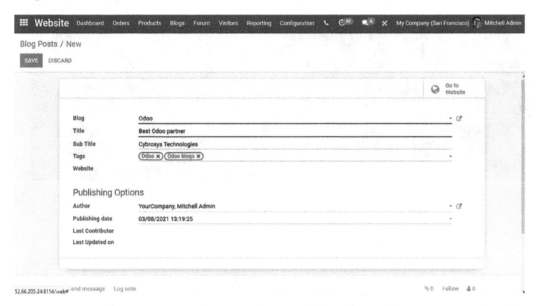

Figure 8.5 – New blog creation window

Once the blog posting configuration is provided and saved, you can select the **Go to website** option available in the creation window, which will direct you to the web page related to the blog post. The following *Figure 8.6* depicts the web page of the blog and, as indicated, the blog is unpublished as it has only just been created. In addition, the name and the subtitle provided for the blog are shown in the window, in the same way as they have been configured in the post creation window. Moreover, you can write the content for the post in the section provided, which can be pasted from a different file, too. The tag names and the other allocations provided are also depicted in the window of the blog, as shown in the following screenshot:

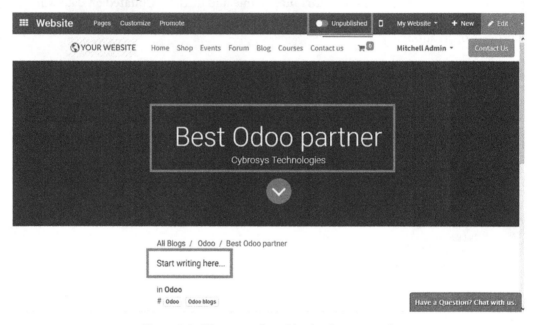

Figure 8.6 – Blog page after a blog has been created

Furthermore, you can configure the various aspects of the blog page using the editor menu available to you after clicking the **Edit** option of the web page in the top-right-hand corner. This will allow you to provide the background, edit the contents, and incorporate various styles, as well as various configuration elements, into the web page, as depicted in the following screenshot:

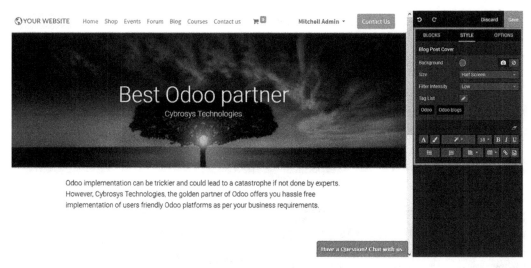

Figure 8.7 – Style editing options for the blogs web page

Once the contents and the web page have been edited and configured, you can save the web page and further publish the blog by selecting the **Unpublished** option and sliding it to **Published**. The blog will then look as shown in the following screenshot:

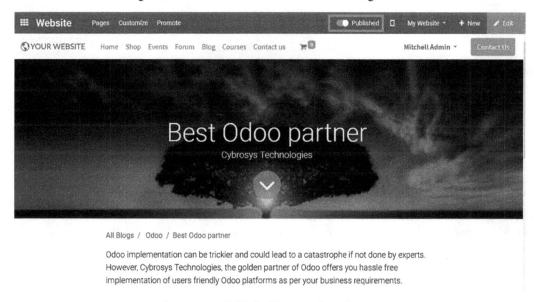

Figure 8.8 – Published blog on the web page

Furthermore, the blog published will be visible on the home screen of the blog's menu, along with the image, blog title, subtitle, tags associated, and the date of publication, as seen in the following screenshot:

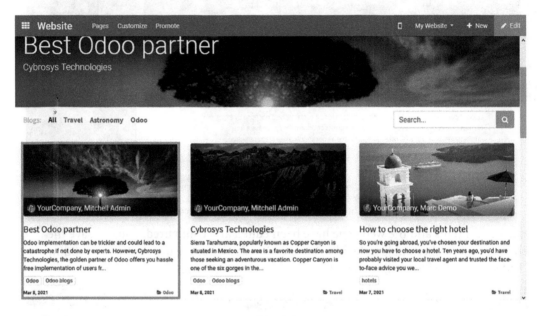

Figure 8.9 – Blog listing menu

Similarly, you will be able to create multiple blog postings on the web page, which is needed and necessary for the company. Moreover, as blog postings are essential marketing tools, required postings can be done regularly to the website using the Odoo website builder. As of now, you will have an understanding of how the blog postings can be created and posted, so let's move on to the design elements of the contents available on the platform.

Designing blog content

The design aspects of the blog content are one of the ways to customize the operations of the blog pages and their aspects for the reader. Moreover, the design of the web page contents can be based on the company's perspective and the standardized design across other web pages, or can stand unique when compared with other website pages. The Odoo website builder recognizes the need for design customization and personalization of the blog pages and provides configurable options, as well as styling elements, to design the blog post web page.

The web page design elements are in the form of the block operations in the Odoo website builder for any page of the website. Similarly, in the case of blog pages, you can use block tools to configure the blog designs. Block tool operation is described in the previous chapter. You can refer to this for a detailed understanding of how to design web pages using the blocks. For your understanding, in the following screenshot, I have added a features block to the web page that will help you to describe the features of an aspect, product, or service to the reader in an attractive manner:

Figure 8.10 – Block tools in the blog page editor

Another aspect related to block design is styling options. This provides you with ample styling options to configure the blocks as well as the content allocated to them. Moreover, the style configuration option will differ based on the blocks and their operations, but the basic configuration options will remain the same throughout the process of creating any block. The following screenshot shows the **STYLE** configuration window of the features block, which will act as the design element for the block operations:

Figure 8.11 – Block style configuration options

Furthermore, there are configuration options that are available in the editing menu. These are block-specific, as well as page-specific, configuration options. You have provisions to set the **Page Layout**, **Background**, **Font Size**, **Font Family**, and **Buttons** configuration options, the border configuration option, and the layout configuration of the web page, as well as various other configuration options, as depicted in the following screenshot:

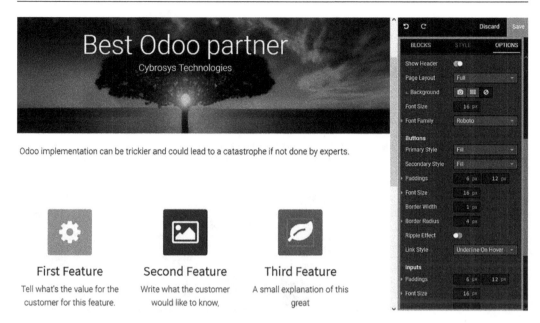

Figure 8.12 – Configuration options in blog page editing

Up till now, we have been discussing the various common configurations and style options available in the Odoo website builder that are generic to any web page operation. Now, let's move on to understand the blog page-specific configuration options.

Blog page customization options

The developers of the Odoo website builder understand the need for a blog page and how important it is to be configured and customized as per company requirements. Therefore, there are certain specific configuration and customization options available for the blog page that will help you to display information or withhold it from being displayed to the web page visitor.

You can select the customize menu available in the web page dashboard. This will present you with various post lists, blog posts, and banner configuration options. These will help you to customize them according to your requirements. Let's look at these options one by one.

Posts List

In the **Posts List** option for configuring the blog page, there are options to enable **Author**, **Comments/Views Stats**, **Cover**, and **Teaser and Tag** posts list options for the blog listing, as displayed in *Figure 8.13*. Enabling the **Author** option will make the platform display the author along with the blog listings. Moreover, enabling the comments or view stats will display the number of both comments and views of the respective blog:

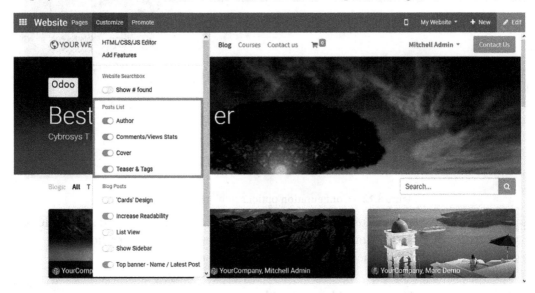

Figure 8.13 – Posts list options in the Customize menu of the blog page

Furthermore, the cover image of the blog that was described will be displayed to viewers on the blog listing page if you enable it under the **Posts List** options. In addition, the teaser and the tags allocated, which provide a preview of the contents and the tags allocated to the blogs, can be enabled or disabled in terms of being displayed to visitors. The resulting web page after enabling all the aforementioned posts list options is as seen in the following screenshot:

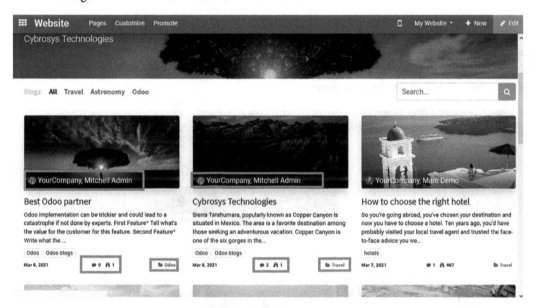

Figure 8.14 – Resultant blog page once the Post List options have been edited

As of now, you will be clear on the Post List options for the blog page. Therefore, let's now move on to the next sections, where the **Blog Posts** options are described.

Blog Post

The **Blog Post** option available in the Odoo website builder blog page editing menu will provide certain illustration options in terms of the web page content, as shown in *Figure 8.15*. You can enable the design of the blog listings to be card-based, which will provide a defined card for each of the blogs. Additionally, there is an **Increase Readability** option for the web page content, which, when enabled, will improve readability for visitors. Moreover, you can change the view of blogs to lists by enabling the **List View** option. In addition, there are options to enable the sidebar and the top banner in the blog listing pages. Furthermore, upon enabling the **Show Sidebar** option, you will see the slide bar configuration options, including options to enable and disable **Archives**, **Follow Us**, and show or hide Tags List on the web page:

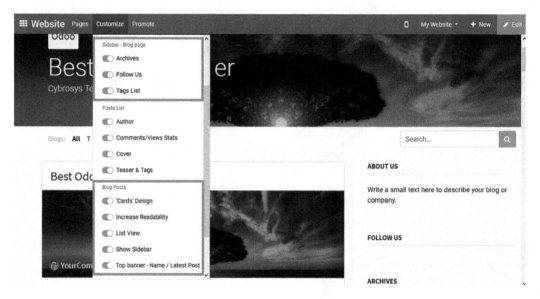

Figure 8.15 – Blog Posts and Sidebar configuration options

The following screenshot shows the resultant web page screenshot once the **Blog Posts** and **Sidebar** options have been configured. Moreover, the **ABOUT US** description, along with the **FOLLOW US** and **ARCHIVES** options of the blog post, is available on the web page as shown:

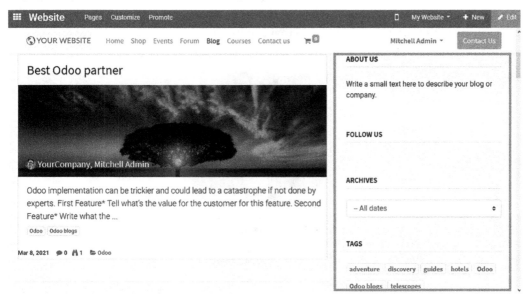

Figure 8.16 – Result of blog post option configurations

You now have an understanding of how blog page editing and configuration can be done using block tools, as well as an understanding of blog page-specific editing and the configuration options themselves. Now, let's move on to understanding the tracking aspects of blog pages.

Visitor tracking

Tracking visitors to the blog is an effective way to understand readers and their activities on the web page, thereby allowing you to create content for a targeted audience in the future. It acts as an effective CRM operation to track visitors, generate leads, and enable the leads generated to result in efficient business opportunities. With the Odoo website builder, you have a visitor tracking option that can be enabled from the customization window, as depicted in the following screenshot:

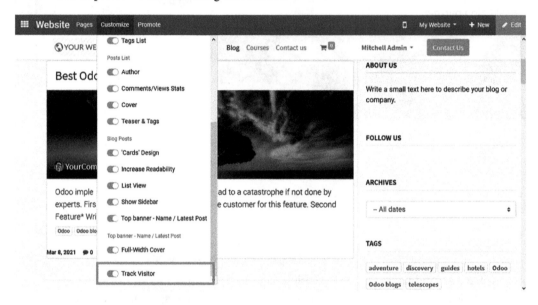

Figure 8.17 – Visitor tracking option in the blog page customization menu

Once the **Track Visitor** option is enabled, the Odoo platform will track visitors to the respective blogs and provide you with a report on that. To understand the visitors who have visited the respective blogs, you can go to the website module and select the **Visitors** tab from the dashboard, which will direct you to the visitor menu shown in *Figure 8.18*. Here, all the blog page visitors will be described to you, along with their status in terms of when they were last active, the number of visits, the pages visited, and the number of pages visited. Moreover, you have provisional options to send out an SMS or an email directly from the page for marketing and promotional purposes:

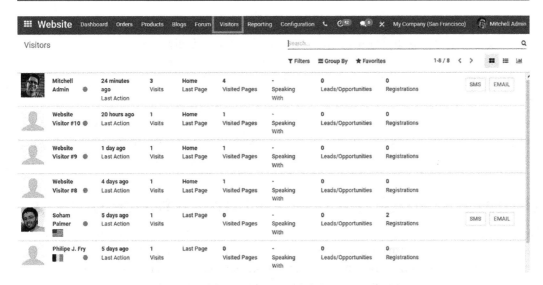

Figure 8.18 – Visitors tab in the backend of the platform

Furthermore, you can select the **Views** tab from the dashboard of the website module under the visitor's tab, which will show the views of the web pages. The **Views** menu will describe the name of the visitors, the pages visited, the URL allocated with the page, and the visit date and time, as depicted in the following *Figure 8.19*. Moreover, there are various default as well as customizable filtering and sorting options available, providing you with ample options to retrieve the required data from a long list:

Visitor	Page	Url	Product	Visit Date
Mitchell Admin		http://52.66.205.24:8114/blog/odoo-3/best-odoo-partner-8		03/08/2021 13:22:04
Mitchell Admin		http://52.66.205.24:8114/blog/travel-1/sierra-tarahumara-1		03/08/2021 13:11:43
Mitchell Admin	Home	http://52.66.205.24:8114/		03/08/2021 12:54:48
Mitchell Admin	Home	http://52.66.205.24:8114/		03/08/2021 11:15:11
Mitchell Admin	Home	http://52.66.205.24:8114/		03/08/2021 10:11:30
Website Visitor #10	Home	http://0.0.0.0:8114/		03/07/2021 18:18:49
Website Visitor #9	Home	http://52.66.205.24:8114/		03/07/2021 02:22:11
Website Visitor #8	Home	http://52.66.205.24:8114/		03/04/2021 23:20:02
Mitchell Admin	Home	http://52.66.205.24:8114/		03/04/2021 14:39:08
Mitchell Admin	Home	http://52.66.205.24:8114/		03/04/2021 11:33:36
Mitchell Admin	Home	http://52.66.205.24:8114/		03/04/2021 11:03:28
Mitchell Admin	Home	http://52.66.205.24:8114/		03/03/2021 16:48:02
Mitchell Admin	Home	http://52.66.205.24:8014/		03/03/2021 10:23:15
Mitchell Admin	Home	http://52.66.205.24:8114/		03/03/2021 10:14:43

Figure 8.19 – The Views menu of the website module

With the help of visitor tracking and the **Views** tab, you can access the list of web page and blog readers to target your future operations and plan accordingly. Moreover, an indirect marketing tool in the form of SMS and email sending options provide you with an option to generate opportunities for business that can be sent out directly from the **Visitors** tab of the platform.

As of now, you have learned about various editing and configuration options available in blog page operations using the Odoo website builder for your website. In the next section, we will be discussing the importance of tags in the blog page of a website.

Understanding the use and importance of tags

Tags have become an important aspect of identification for content in the websites. Today, we assign tags for blogs, images, web page constants, products, and many other things besides. Moreover, tags have become a classification aspect and filtering parameter in a long list of data available to you. In a web page or a website, the tags for content, as well as images, videos, or blogs, are allocated by the web page developer or the SEO manager in charge of the website.

The Odoo website builder gives you the ability to assign and provide tags to the web page contents. Moreover, assigning can be done with ease and you can create several tags for your website, company specifications, and requirements. Furthermore, considering blog pages, the tag functionality remains an important aspect of classification. The tags remain as a simplifier for search functions, since when a visitor searches for a specific blog, they can search the long list of blogs based on the tags. Furthermore, you can filter out the required tags, providing just the necessary information as well as the blogs with the assigned tags.

The **Blog Tags** window for the website can be seen in the Odoo website builder under the website module and by selecting the **Blog Tags** option from the **Configuration** tab. Here, all the blog tags that have been described and created are depicted, along with the number of posts they have been allocated to. The following screenshot shows the **Blog Tags** window of the website module:

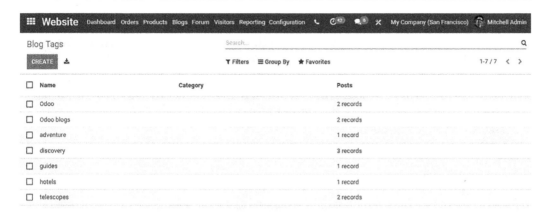

Figure 8.20 – Blog Tags window of the website module

Moreover, you can create blog tags by selecting the **Create** option, which will direct you to the window depicted in *Figure 8.21*. Here, you can provide a name to the tag and assign the category to which the respective tag is allocated. In addition, you have the provision to create and edit a category directly from this window. Here, we have provided the tag name as New, and the category as Test. You can also specify the tags to be used in the various blogs described in the website under the **Used in** section, as depicted in the following screenshot:

Figure 8.21 – New blog tag creation window

Select the **Add a line** option available and you will be presented with a list of all the blogs described on the website. When the **Add a line** option is selected, the **Add: Posts** window (see *Figure 8.22*) will pop up. Here, all the blog lists will be described, along with the author, the number of views, last updated details, blog type, and many more details besides. Furthermore, you can select the tick box available relating to each blog for the tag to be assigned to them and click on the **Select** button at the bottom of the window to add these blogs to the **Blog Tags** window:

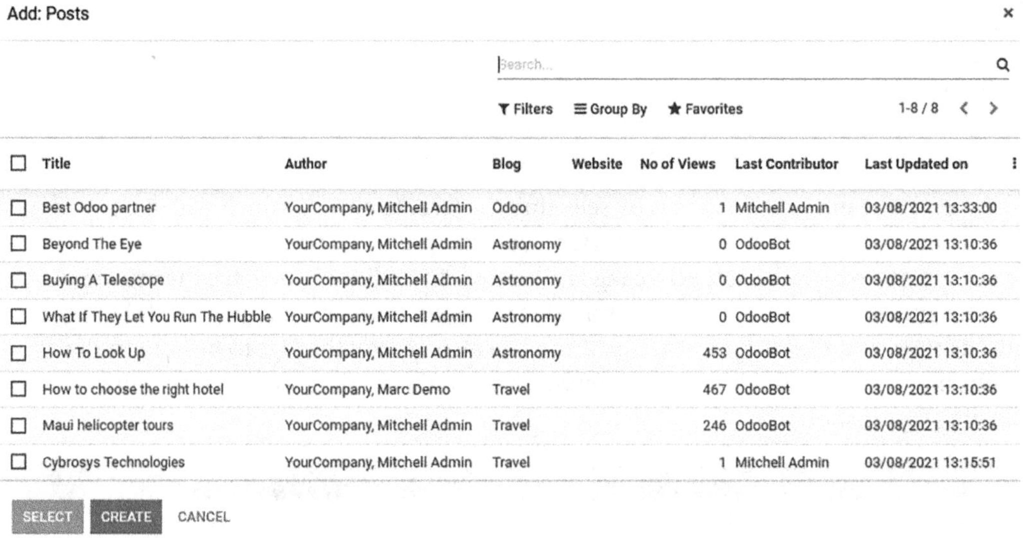

Add: Posts   ×

	Title	Author	Blog	Website	No of Views	Last Contributor	Last Updated on	
☐	Best Odoo partner	YourCompany, Mitchell Admin	Odoo		1	Mitchell Admin	03/08/2021 13:33:00	
☐	Beyond The Eye	YourCompany, Mitchell Admin	Astronomy		0	OdooBot	03/08/2021 13:10:36	
☐	Buying A Telescope	YourCompany, Mitchell Admin	Astronomy		0	OdooBot	03/08/2021 13:10:36	
☐	What If They Let You Run The Hubble	YourCompany, Mitchell Admin	Astronomy		0	OdooBot	03/08/2021 13:10:36	
☐	How To Look Up	YourCompany, Mitchell Admin	Astronomy		453	OdooBot	03/08/2021 13:10:36	
☐	How to choose the right hotel	YourCompany, Marc Demo	Travel		467	OdooBot	03/08/2021 13:10:36	
☐	Maui helicopter tours	YourCompany, Mitchell Admin	Travel		246	OdooBot	03/08/2021 13:10:36	
☐	Cybrosys Technologies	YourCompany, Mitchell Admin	Travel		1	Mitchell Admin	03/08/2021 13:15:51	

SELECT   CREATE   CANCEL

Figure 8.22 – Pop-up window for adding blog posts allocated to the tags

Now, coming back to the blog web page of the website, you can view the assigned tag on the respective blog as depicted in *Figure 8.23*. This will only be visible if the **Teaser and Tags** option has been enabled from the **Customization** window, as discussed in previous sections of the chapter:

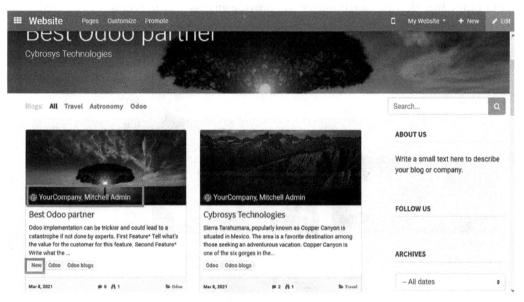

Figure 8.23 – Resultant blog page following blog tag allocation

Additionally, the visitor also has the option to search the blogs based on the tags by providing the tag name in the search field available on the **Blogs** page of the website, as shown in *Figure 8.24*. However, this will not be beneficial for new visitors to the blog pages as they will not be aware of the tags allocated to the blogs. To a frequent visitor and reader of your blogs, this filtering functionality will provide an easy way to access the required blogs and their categories:

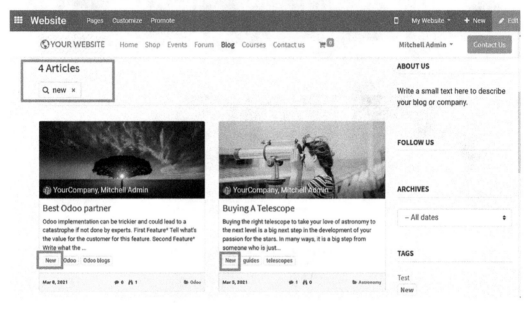

Figure 8.24 – Search results based on blog tags

Now you know how tags can be created and assigned to the various blogs that will be created, and those that have already been created. Let's now move on to understand the tag categories of the Odoo website builder and how they can be created and managed.

Tag categories

Tag categories are regarded as tag classification elements in the Odoo website builder. However, these tag categories are for recognition and in-house operations of the company on the blogs of the website and are now visible and available to visitors to the web pages for filtering out blogs or tags, as mentioned in relation to the creation of tags in the previous section.

The **Tag Category** window can be accessed from the website module under the **Configuration** tab. In the menu, all the tag categories created to be operational will be visible. Moreover, you have the option to download the list to export it to a different platform. The following screenshot shows the **Tag Category** window of the website module:

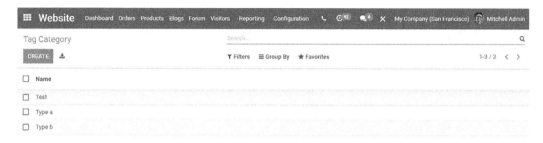

Figure 8.25 – Tag Category window in the Odoo website module

In addition, you have the ability to create new tag categories using the **CREATE** option available. When you select the **CREATE** option, the following window in *Figure 8.26* will show up. Here, the new tag category **Name** can be provided and created after saving it:

Figure 8.26 – Tag category creation window

Upon creating a tag category, you will be able to assign the various tags to the respective category while creating a new tag in the tag creation window.

Now that you know how the tags and the tags category can be created and managed, let's now move on to the next section where the SEO management aspects of the Odoo website builder will be discussed.

SEO management

SEO is one of the most important aspects of website operations. The people working in the SEO teams try to make sure that your website will be discovered more easily to drive more visitors to it. In addition, they play a crucial role in showcasing website content to bring in more web page traffic. Moreover, the Odoo website builder provides you with dedicated SEO optimization tools for the promotion and boosting of your web pages. These tools are user-friendly as regards operation and you can become an expert in using them in next to no time.

In addition, the Odoo website builder has a dedicated menu for the SEO aspects of each web page. You can select a web page from your website and navigate to the **Promote** menu available on the dashboard. You will then be presented with the **Optimize SEO** option, as seen in the following screenshot:

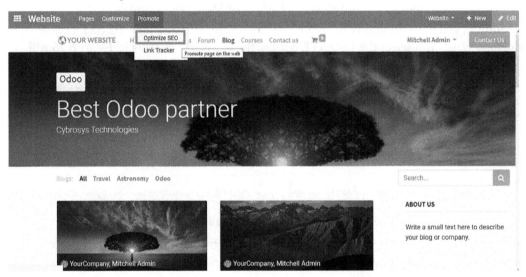

Figure 8.27 – The Optimize SEO option in the Promote menu

Upon selecting the **Optimize SEO** option, you will see the **Optimize SEO** window, as shown in *Figure 8.28*. Here, you can provide the title for the web page. Moreover, the title of the web page will be displayed, which can be modified in the Title option. Furthermore, a description of the web page can be provided, which will be displayed to the visitor upon serving the web page. Moreover, the preview of the page title and the description allocated can be visualized in the preview available on the right-hand side, as depicted in the following screenshot:

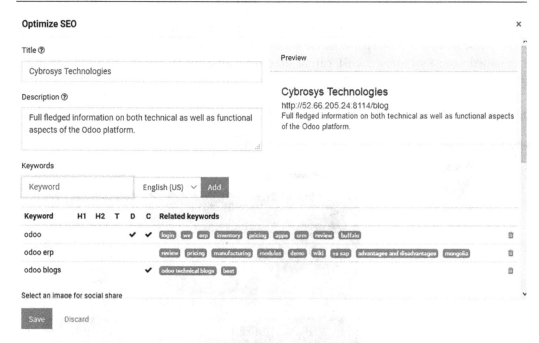

Figure 8.28 – The Optimize SEO window

Furthermore, you have the provision to add keywords based on content for your web page to be visible when searched for by customers. The keywords allocated here will determine whether your web page should be shown when searching for a specific aspect. You can add *n* number of keywords depending on your needs. However, it is advisable to use one genuinely relevant keyword rather than allocating a bunch of keywords in terms of SEO.

In addition, you can configure the images that will be shown in the event of a web page search. These images will be visible in the various social share operations. The images can be uploaded from a server or a URL can be used from an external website or web page, as shown in the following screenshot:

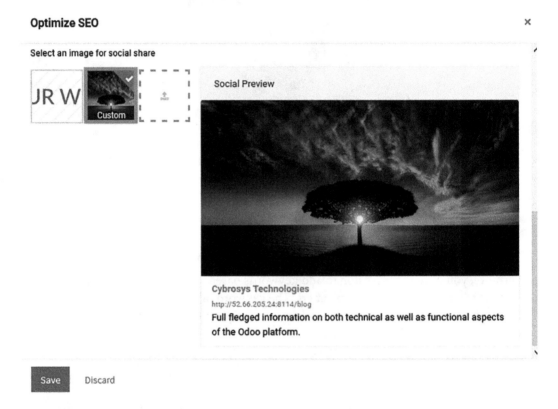

Figure 8.29 – Image configuration options for social share

With the various SEO tools available, you can be sure that you'll have a high amount of web page traffic to your websites, which will ensure that you have numerous blog readers for your website blogs that are published using the Odoo website builder.

Today, blogs have become an essential part of website operations and they remain one of the best forms of content writing to attract readers as well as customers. Moreover, these blogs will act as a perfect marketing and promotional tool if exercised in a precise manner. With the Odoo website builder, you will be able to create, publish, and customize the blog web pages of your website.

Summary

The Odoo website builder has ample options, including blog creation tools, publishing options, and tag allocation, along with an SEO optimization menu, which will act as reliable tools for creating and managing the blog web page of your website. You now have a basic understanding of and the ability to create, publish, and manage your web page of blogs using the Odoo website builder.

In the next chapter, we will be discussing how you can go live with your e-commerce website using the Odoo website builder and its allocated tools.

Questions

1. How can you classify the blog tags in the Odoo website builder?
2. Where can you create and allocate tags for blogs in Odoo?
3. How can you provide a web page description to be shown in search results?

Further reading

- *Working with Odoo*, by Greg Moss, Packt Publishing
- *Learning Odoo*, by Greg Moss, Packt Publishing

9
Go Live with Your E-Commerce Website

In the previous chapter, we discussed how to **create blogs** for your website, add a web page to it, and how to manage them. Moreover, we learned about the **SEO optimization** aspects of the Odoo website builder. Now we will move on to understand how you can define and run e-commerce operations in the Odoo website builder. Furthermore, we will be focusing on the creation, design, manageability, and use of various capable tools in Odoo for e-commerce operations.

In this chapter, we will be focusing on the **e-commerce website** aspects of building using the Odoo website builder. To do so, we will cover the following topics:

- Adding a product to an e-commerce website
- Functional options for an e-commerce web page
- Designing an e-commerce page
- Adding blocks to an e-commerce page
- Adding a carousel to e-commerce page

- Creating a separate page for product categories
- Configuring different payment acquirers for customer payments

By the end of this chapter, you will be able to create and design an e-commerce website for your company using the Odoo website builder tool.

Technical requirements

A prerequisite basic knowledge of e-commerce website operations along with their functioning in a live environment will be an added advantage to learn the various aspects of building such a site using the Odoo website builder. Furthermore, an insight into the Odoo website builder along with an understanding of the various options and features involved would be beneficial. In addition, a system with a moderate processing speed and the Odoo platform installed is vital.

Adding a product to an e-commerce website

E-commerce websites and platforms have become the modern way of doing business rather than setting up a retail shop with limited accessibility. This is considered one of the major advantages of e-commerce platforms over conventional ways of doing business. You have no limitations in expanding the reachability of products to new areas, which can be done with ease and the same amount of investment. Another aspect to consider is the low amount of investment and facility requirements when compared to a retail as well as a wholesale business.

In addition, with e-commerce platforms, you can add *n* number of products to be sold with the same amount of investment, using which the entire operations are performed. You can add and remove products based on credibility, business statistics, and various other business reasons. With the Odoo website builder, you can add a product or remove one with ease. The operations regarding adding as well as removing or managing products are done from both the frontend and the backend of the platform and you only need to have functional knowledge of Odoo website builder operations. Moreover, the operations are well defined in the Website module of the platform and can be done with ease.

To add a product to an e-commerce website using the Odoo website builder, you first need to log into the website. Now, from the relevant web page, select the **New** option as shown in *Figure 9.1*.

The option is available from all websites if the website editing operations are being conducted by company officials. Note that the option is not visible to web page visitors and other unauthorized users of the website:

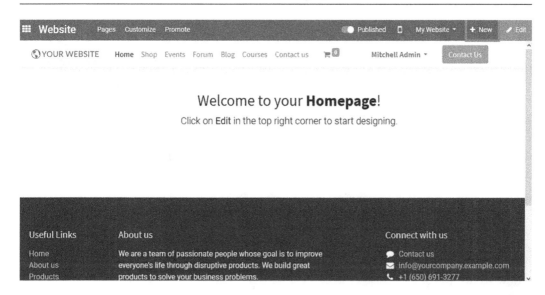

Figure 9.1 – The New option on a web page of your website for editing it

Upon clicking the **New** option, you will be taken to the menu to add new components to the website of the company. You have the options to add a **Page**, **Blog Post**, **Event**, **Forum**, **Job Offer**, **Product**, **Course**, or **Appointment Form**. From the menu, select the **Product** icon (see *Figure 9.2*):

Figure 9.2 – Menu to add new components to the website

Upon selecting the **Product** icon, you will be shown the pop-up window seen in *Figure 9.3*. Here, you can initially add the product **Name** and opt for the **Continue** option to add the product to the e-commerce website. Moreover, you can provide any customized product name for the product and the name provided here can be further edited:

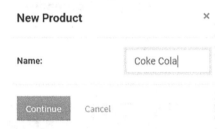

Figure 9.3 – The pop-up window to add a new product

Once you click on the **Continue** button, you will be directed to the product description window shown in *Figure 9.4*:

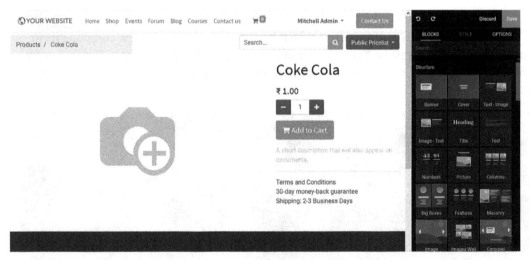

Figure 9.4 – The product description window of the new product along with the web page editing menu

Here, the product name will be what you added in the **New Product** window. In addition, the **Terms and Conditions** will be as described by the company policy and can be further edited based on your requirements. You can also change the price of the product from the original one.

Next, to change or add an image, select the image icon and you will see an upload image window from which you can either add the URL or the image available on the desktop of the system. Moreover, there are image **Filter** as well as **Width** configuration options available, which will help you to configure the style of the image.

In addition, with the **Quality** configurator, you will be able to manage the image resolution and clarity aspects, as shown in the following screenshot:

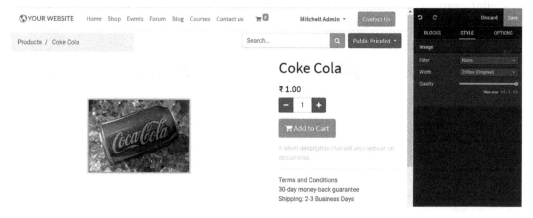

Figure 9.5 – Style configuration menu for the product

You can also add various blocks to the product description to provide details. Selecting the **BLOCKS** menu from the editor window will show you all the block types of the Odoo website builder. Moreover, you can select a block by dragging and dropping it on the web page to use it, just as depicted in the following screenshot:

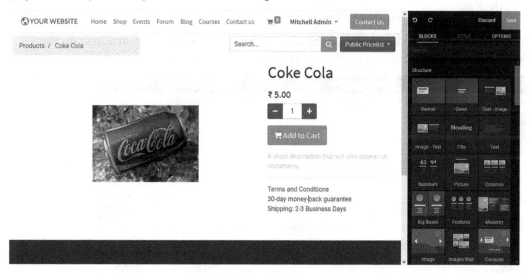

Figure 9.6 – The BLOCKS menu for web page editing

Up until now, we have been discussing aspects of adding a product from the frontend of the platform. Now, let's move on to understanding how it can be done from the backend of the Odoo platform.

To add a product to the company website, select the **Website** module and select the **Products** menu from the dashboard. The following screenshot depicts the **Products** menu where all the products are described with the **Product Name**, **Internal Reference** details, the **Responsible** person, **Sales Price**, **Cost**, **Quantity on Hand**, **Forecasted Quantity**, and the **Unit of Measure** details:

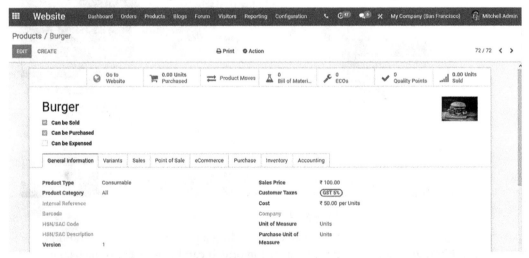

Figure 9.7 – Products menu in the Website module

You can add a new product to the list by choosing the **CREATE** option available on the **Products** page and providing the details in the menu as shown in *Figure 9.8*:

Figure 9.8 – Product creation window

In the product creation window, you can provide the name of the product, assign an image, give all the **General Information** on the product, provide the product **Variants**, information give the **Sales** information on the product, provide the **Point of Sale** details on the product, and the **Purchase** details of the product can be mentioned along with **Inventory** and **Accounting** details.

To add the product description that will be presented on the e-commerce website, you can select the **eCommerce** tab available in the window. In this tab, you can assign the website on which the product should be described if you are operating the Odoo platform to be functioning with multiple websites. In addition to this, you can assign the **Categories**, **Alternative Products**, and **Accessory Products** information for the main product that should be described on the web page of the main product that you have created in the options, as shown in the following screenshot:

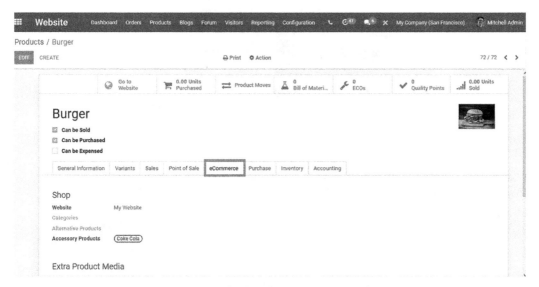

Figure 9.9 – Menu to edit the e-commerce details of the product

Upon providing the details, you can select the **Go to Website** option available on the product creation page, which will direct you to the product page on the e-commerce website and you can view the product being described and see the way you described it in the backend of the platform. For the product to be published on the platform, you can select the **Unpublished** option and slide it to **Published** on the e-commerce website as shown in the following figure:

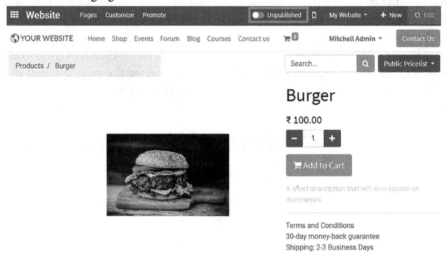

Figure 9.10 – Product description on the e-commerce page of the website

In addition to this, you have the option to add a block to the web page just as described before, although the product is added from the backend of the platform and is as depicted in the following screenshot:

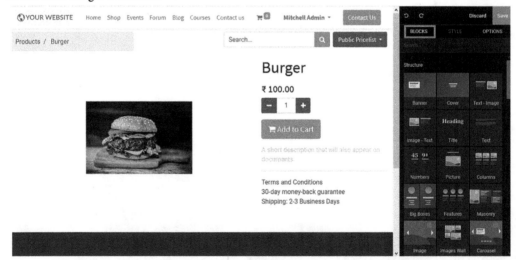

Figure 9.11 – BLOCKS editing menu in the product description window

Moreover, all the **STYLE** configuration, as well as editing **OPTIONS** available on the platform are the same as the ones in the previous chapters where the types of blocks and their operations are described.

With the option to add the product from both the frontend and the backend of Odoo website builder operations, you can configure and describe the product the way you want. Moreover, any product added here will be available for management through the Inventory module and will be shown in the Sales module and described in the purchase module to deal with all the related operations on the product.

Ribbons

The ribbons of the product displayed are one of the best ways to communicate the status of the product based on the company inventory. The Odoo platform permits the use of ribbons based on the availability status of **Sold out**, **Available**, or **New**. This can be configured under each product that is being described on the platform. The following screenshot depicts the ribbon configuration options, which are highlighted for **Cable Management Box**:

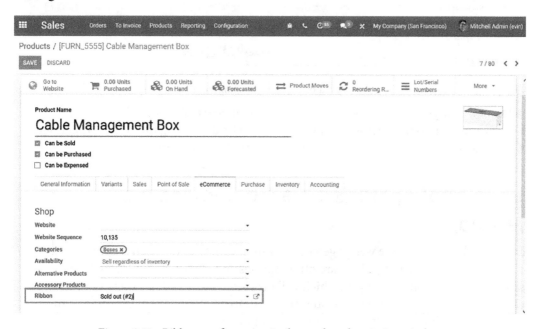

Figure 9.12– Ribbon configuration in the product description window

Now, once the ribbon is configured and the customer logs into the company website or the e-commerce platform, the respective product will be depicted with the ribbon described, as shown in the following screenshot:

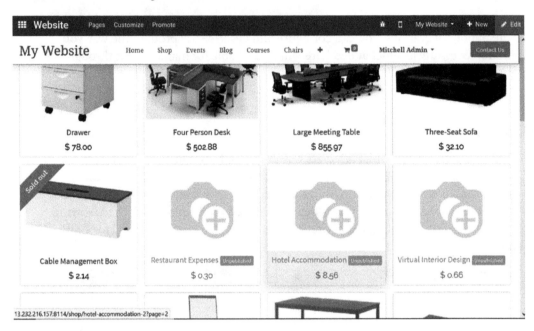

Figure 9.13 – Ribbon on a product in the Shop menu of the website

You learned how products can be added to an e-commerce platform in this section. Let's now move on to the next section, where we will be discussing the various functional options available for an e-commerce page run on the Odoo platform.

Functional options on an e-commerce web page

Odoo provides e-commerce web page visitors with a variety of functional options that can be configured. Moreover, there are other options as well, which can be enabled and disabled from the frontend by you, which will provide visitors with a variety of functional tools while searching for products. Under the **Customize** menu, just as depicted in the following screenshot, you have these functional options visible:

Figure 9.14 – Functional editing options on an e-commerce web page

Here, you can enable the **Show # found** option, which falls under the **Website Searchbox** options for the products that are being searched based on the respective hashtag being made visible. Moreover, there are various **Products Item** options available, such as **Comparison List**, **Product Description**, and **Wishlist Button**, which can be enabled by the visitor, as well as yourself, from the menu.

Additionally, there are various **Products** options available, such as **Grid or List button**, which will show the options for visitors to choose whether products should be shown as a grid view or list view. You can also enable the **Images Full** option for product images to be depicted at full scale. Moreover, the **List view** option can be enabled for it to the default view of the products that have been described. In addition, you can enable or disable the **Product Attribute's Filters** option. Furthermore, you can enable the **Show Sort by** option for you to configure the product sorting operations for visitors. Additionally, **eCommerce Categories** will be viewable to visitors by enabling the option available. In the next section, we will be discussing the design aspects of an e-commerce website.

Designing your e-commerce page

The design aspects of e-commerce web pages play a crucial role in the appearance of products as well as the menu items present on your web page, along with the various options available. If you have visited various e-commerce websites, you might have observed their eye-catching design, the alignment of their content, and attractive product descriptions depicted in a format that is easily readable and accessible with all the required information.

With the Odoo website builder, the design elements of e-commerce websites can be taken care of in all aspects of e-commerce web pages. As seen in all other web page operations of the Odoo builder, you have the option of generalized block tools available here as well. Besides the block tools, there are various other styles and configurational options available as well, which will provide you with ample options to design and configure an e-commerce web page.

Style options

Under the **STYLE** configuration menu, you can provide the design elements to the header section as well as the navigation bar of the web page. There are various options under the **Header** section. You can set a **Template** from the various default ones available by choosing the drop-down option. Now, as the template is set, you have the provision to set **Colors** for the header, which can be chosen from a wide-ranging selection. Furthermore, the **Border** width size state and the color of the borderlines can be chosen similarly. Style configuration options such as **Round Corners**, the **Shadow** of icons, **Color**, **Offset (X, Y)**, **Blur** effect, **Speed** effect, **Scroll Effect**, and the option to enable and disable **Show Empty Cart** can be chosen as per the requirement. The header style configuration options are depicted in the following screenshot:

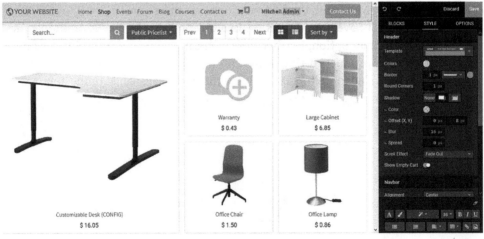

Figure 9.15 – Style configuration menu for an e-commerce page

Now let's move on to the options available in the navigation bar style configuration options in the **STYLE** menu as shown in the following figure:

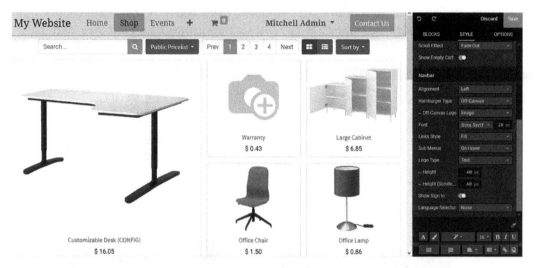

Figure 9.16 – Navigation bar editing option for an e-commerce web page

Here, you have configuration options to set up the **Alignment**, **Hamburger Type** for the listing of icons, an **Off-Canvas Logo** that image can be loaded or added, **Font** style and size, and the **Links Style** and **Sub Menus** that can be added. Furthermore, you have options to configure **Logo Type**, the **Height** of the navigation bar, the **Height** of the scroller, and the option to enable or disable **Show Sign In** for the user is available. Additionally, if there is a requirement for **Language Selector**, it is also available and can be chosen.

Configurable options

The web page editing menu of the Odoo website builder has an **Options** menu where the various options to configure a web page design and associated elements can be set. Here, you have to choose the **Theme colors** of the web page for the various default styles available, which can be further edited. Furthermore, there are various **Theme Options** available, which will be described in the following subsection.

Theme Options

The theme editing options available in the Odoo website builder will provide you with the appearance configuration options to design your web page. You have options to enable and disable **Show Header**, describe **Page Layout**, set **Background**, set **Font Size**, and choose a **Font Family** option, as seen in the following screenshot:

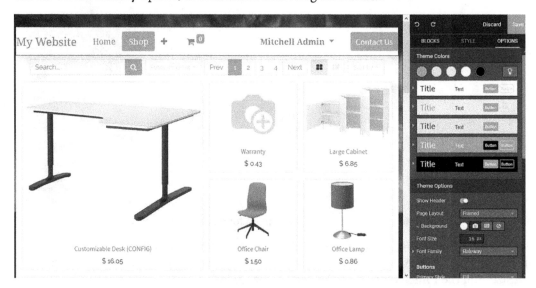

Figure 9.17 – Configuration options for e-commerce page editing

The theme options in the editing menu will help you to configure and describe your web page as you per your needs.

Buttons options

The next type of configuration option available under the editing menu is the **Buttons** configuration options. Here you can configure the primary style as well as the secondary style of buttons. Furthermore, **Paddings**, **Font Size**, **Border Width**, and **Border Radius** can be configured. In addition, you can configure the **Link Style** of the link being provided and enable or disable **Ripple Effect** on the web page as shown in *Figure 9.15*.

Inputs options

The next set of configurable options available in the web page editing menu of the e-commerce page is the **Inputs** option. Here, you can configure the **Paddings**, **Font Size**, **Border Width**, **Border Radius**, and **Status Colors** settings of the web page as seen next:

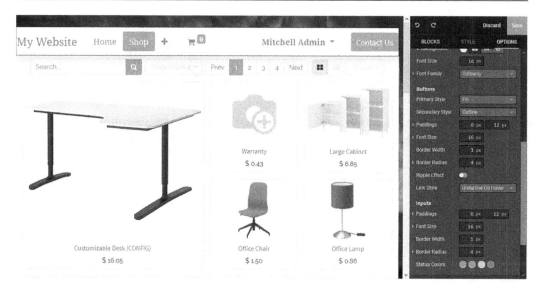

Figure 9.18 – Buttons and Inputs configuration options

With all the configurable options available in the existing menu, you will be able to design and set up the web page as per your requirements for your required company operations. Let's now move on to the next section, where blocks on e-commerce web pages will be described.

Blocks on your e-commerce page

The block operations of the Odoo website builder showcase one of the unique features of designing a web page with ease. Block tools can be easily dragged and dropped onto web pages. The block operation for the e-commerce web page and platform operating with the Odoo platform has similar block tools to define the contents and design style elements in the aspect of appearance.

As seen earlier, the block tools can be easily selected, dragged, and dropped to the desired web page area of the e-commerce, which will help you to describe the contents. Moreover, you will have various block configuration and style editing tools that will provide you with further operational capabilities and functional options to design the web page. Let's move on to discuss the various types of blocks available in the Odoo website builder for e-commerce page editing.

The Structure block

The first type of block tool available in Odoo is the **Structure** block. This helps bring an initial structure to the description of contents on the web page. Under the **Structure** block classification, there are many styles of block tools available, from which you can choose. In total, there are 18 types of Structure blocks available, such as **Banner, Cover, Text-Image, Image-Text, Title, Text, Numbers,** and many more. The following screenshot shows the editing menu where the Structure block is described:

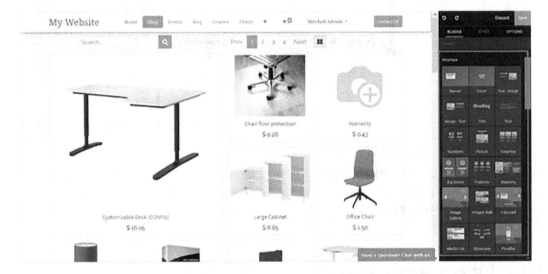

Figure 9.19 – Structure block in the BLOCKS tool menu

The description of the various structure blocks and the aspects of their operations have been previously described in the book – you can refer to *Chapter 3, Introduction to Blocks – Structure Blocks*, where the Structure blocks are well described and explained one by one. The operations of a Structure block on an e-commerce web page are similar to those described on all other web pages of the Odoo website builder. Let's now move on to understanding the Features block of the Odoo website builder in the next section.

The Features block

The **Features** block in Odoo is another block tool classification that will help you to describe the contents of a web page with the help of the various block tools available under the **Features** block. In total, there are around 13 types of **Features** block tools available under the classification, which will help you to provide the desirable contents block tools such as **Comparisons**, **Team**, **Call to Action**, **References**, **Accordion**, and many more block tools under the **Features** block classification. The following screenshot shows the web page editing menu and the **Features** block listing in it:

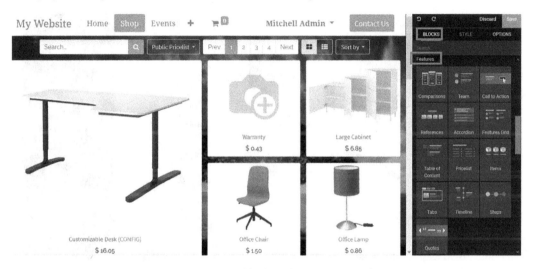

Figure 9.20 – Features block in the BLOCKS tool menu

The **Features** block and the various tools in operation were described previously in the book (you can refer to *Chapter 4, Design Using Features Blocks*, where the **Features** blocks are well described and are explained one by one in detail). The operations of these blocks on e-commerce web pages are similar to the usage of them on all other web pages. Moreover, the block will provide a designated area on the web page to describe the various distinctive aspects of the company and its products and services. Let's now move on to understand the **Dynamic Content** block in the next section.

Dynamic Content

The **Dynamic Content** block is yet another type of block tool available in Odoo, which will provide you with description aspects on web pages. Moreover, using these block tools, you will be able to describe various contents for company websites in detail. There are 16 different types of Dynamic Content blocks available in the Odoo website builder, such as **Forms**, **Google Map**, **Dynamic Products**, **Viewed Products**, and many more to describe web page contents. The following screenshot will take you through the dynamic blocks described in the editing menu of the Odoo website builder:

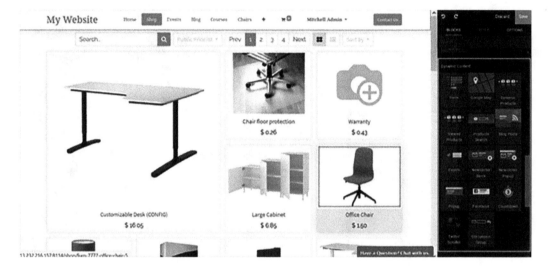

Figure 9.21 – Dynamic Content block in the BLOCKS tool menu

The Dynamic Content block of the Odoo website builder is well described in *Chapter 5*, *Designing a Website using Dynamic Content* for a quick recap. The Dynamic Content block operations of the e-commerce web page are similar to the block operations on all other web pages. As of now, you will have an understanding of the Dynamic Content block, therefore, let's now move on to the next section, where the **Inner Content** block is explained.

The Inner Content block

The **Inner Content** block is the last type of block classification available in the Odoo website builder. These block tools are used to describe the content of web pages with a defined appearance and style. In addition to the various styles and configuration options available for the respective block, you will be able to describe the contents as per your requirements. There are 13 Inner Content block types available, such as **Separator**, **Alert**, **Rating**, **Card**, **Share**, and many more blocks for defining the contents of a web page:

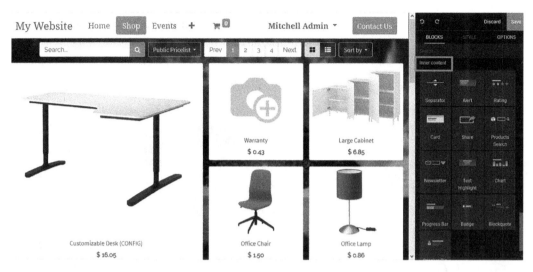

Figure 9.22 – Inner content block in the BLOCKS tool menu

The Inner Content blocks are well described in one of the previous chapters of the book, which explains aspects of how the various blocks under this classification can be used and be made functional as per your web page requirements. You can refer to *Chapter 6, Inner Content Block Tools*, to have a recap of all the operations and functional aspects of the Inner Content blocks.

The block tools of the Odoo website builder can play a crucial role in the design aspects of e-commerce web pages, which can bring in attractive forms of content as well as appearance aspects for e-commerce platforms. Till now, we have been discussing the block tools available in the Odoo website builder. Let's now move on to learn how to add a carousel to an e-commerce web page of a company website.

Adding a carousel to your e-commerce page

A carousel on an e-commerce page can bring wonders to the web page as you will be able to describe specific features of products or services you offer and give them catchy titles in each carousel slide. A carousel for an e-commerce web page can be added in the Odoo website builder from the block operations in the editing menu. The **Carousel** block comes under the category of Structure blocks of the Odoo website builder. You can simply drag and drop it to an e-commerce web page at the top or the bottom of the page, after the header and before the footer of the web page respectively.

There will be default content provided to the **Carousel** block that you have selected, and you can edit the contents as per your requirement. Furthermore, there are various style configuration options available for the **Carousel** block, just as depicted in the following screenshot:

Figure 9.23 – Style configuration options for the Carousel block

You can set the **Background** for the carousel block to be an image, video, or clipart of a shape that can be changed, replaced, and removed whenever you require it. There are further configuration options to set **Position** and **Parallax** for the background. In addition, you can configure the **Style**, **Transition**, **Speed**, and **Height** fields for the **Carousel** block that is being created. Furthermore, you can enable or disable the **Scroll Down Button** setting.

Coming to the **Slide** configurations of the **Carousel** block, you can add or remove any number of slides from the block. A slide can be selected to change its **Background** to an image, video, graphics, or a shape that can be replaced and changed. Furthermore, just as for the block styling operations, you have **Position**, **Parallax**, **Filter**, **Width**, **Quality**, and **Content Width** configuration options for each slide, as depicted in the following screenshot, which can be set to suit any design requirement for each slide as shown in the following screenshot:

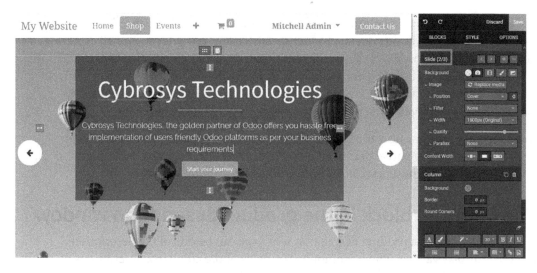

Figure 9.24 – Slide editing under the Carousel block

Upon editing the slides and the block content, you can save the **Carousel** block to be operations on the web page just as seen in *Figure 9.22*. The visitor will be provided with a slider where the slides can be moved manually or can be automated by setting a time interval, which will be as described in the various configuration options mentioned previously:

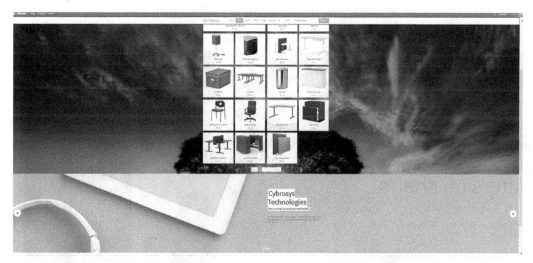

Figure 9.25 – Resultant web page after the Carousel block is inserted

A **Carousel** block for an e-commerce website will be an eye-catching and informative tool that can convey various offers, product specifications, advantages, and all aspects of services as well as products. You might have visited various shopping websites and viewed **Carousel** blocks being one of the informative ways to display the offers available on various products, just as the Odoo website builder will allow you to configure the **Carousel** block for application-specific operations as per your requirement.

In this section, you saw how the **Carousel** block can be used on an e-commerce web page to act as an informative part of an e-commerce website. So, let's now move on to understand how a carousel can be set on a product description window.

The Carousel block in the product description window

A **Carousel** block on a product description web page can be one of the description tools that display various offers, promotional aspects of the company, descriptions of promotional and marketing aspects, and can be used to provide an option to direct visitors to a different web page of an e-commerce website. Let's look at how you can add a carousel to a product description web page.

On the web page, select the **Edit** option, choose the **Carousel** block, and drag and drop it to the respective location on the web page. Provide the details and edit the required content as per the requirement. You can configure the options available in the **Carousel** block as depicted in the following screenshot, by double-clicking on it:

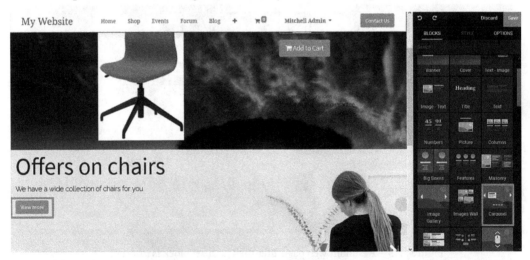

Figure 9.26 – Adding a Carousel block to the product description window

Upon double-clicking on the option, a pop-up window will come up, as seen in the following screenshot:

Link to   ✕

URL or Email   /shop/category/furnitures-chairs-11   Preview

Hint: Type '/' to search an existing page and '#' to link to an anchor.

Type   Link ✔ Primary   Secondary

      Open in new window

Save   Discard

Figure 9.27 – Attach the link to the menu for the Carousel block button

Here, you can provide the **URL or Email** address of the web page the visitor should be directed to. Here, we provide a link to the product category menu of the e-commerce web page. Upon configuring the button, you can click on the **Save** option in the window.

Now, when the visitor logs into the web page and clicks on the respective carousel button from the product description menu, they will be directed to the product category web page as shown in the following screenshot, which has been pre-defined. Moreover, the link to the button can be provided to the contact page of the company, or a form can be shown for visitors to fill out to enquire about products and services:

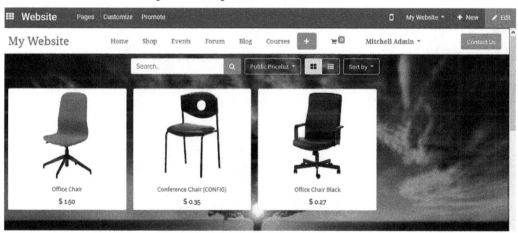

Figure 9.28 – The product category window the visitors are directed to through the button in the Carousel block

The **Carousel** block of the Odoo website builder is one of the tools that can be well used in the e-commerce website operations of a company. With the Odoo website builder, you will be able to configure and describe the carousel lock for various specific operations.

Let's move on to the next section where we will describe how to create a separate page for product categories on an e-commerce website.

Creating a separate page for product categories

Assigning a product to a category will be a beneficial operation for categorizing the product for the in-house operations of the company. For the visitor, it will be the best tool for filtering out products. Moreover, the product category description can be done from the backend of the platform when products are described in the inventory module. On the e-commerce website, you will be able to add a separate page for a product category or one for all the product categories as it will help visitors to search for a product based on the category classification.

For the product categories to be added to the website of the company, you should initially create a product category in the **Inventory** module of the platform. To do so, select the **Product category** menu available in the **Configuration** tab of the module and you will be shown the menu where all the available **Product Categories** are described, just as depicted in the following screenshot. Here, you can edit a product category or select a new one to be created. As we already have a category, **All / Chairs**, we will move on to edit it as seen in the next figure:

Figure 9.29 – The Product Categories menu in the Odoo website module

In the description window, the category name, in our case **Chairs**, **Logistics** information, **Inventory Valuation** details, and **Account Properties** will be described, which can be edited by selecting the **EDIT** option available in the **Product Categories** menu. While creating a product category, you should define all these aspects. The product under the respective product category will be described in the **Products** menu seen at the top-right of the web page, just as seen in the following screenshot:

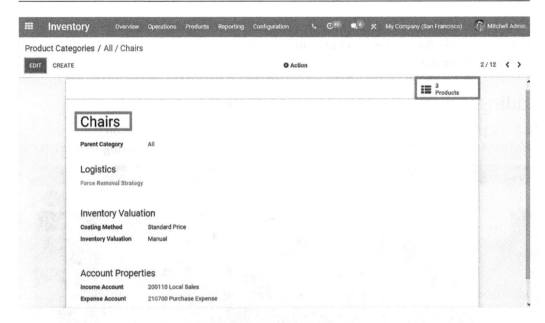

Figure 9.30 – Product Categories creation menu

The **Products** menu of the respective product category will list out all the products under it and you have provisions to create a new category by selecting a product and changing the product category to the one required. Once the product category is defined, you will be able to filter the products based on it or add a web page to the e-commerce website for the respective product category:

Figure 9.31 – Products under the Chairs product category

Up until now, we have been discussing the configuration of product categories on the backend of the platform. Now let's move on to see how to add a product category page to the website in the next section.

Adding a product category page to the website

To do so, you can go into the website and select the **Pages** option available on the dashboard and then select the **Edit Menu** option, just as seen in the following screenshot:

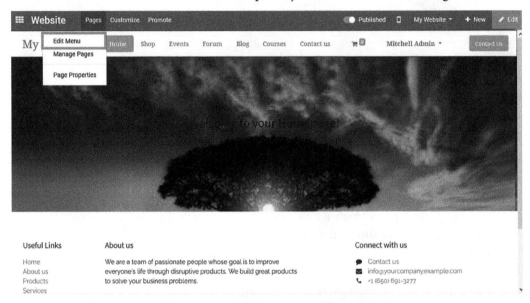

Figure 9.32 – Page edit option on the website

Upon choosing the **Edit Menu** option, you will see an **Edit Menu** pop-up window. Here, all the menus on the website are described and you have the provision to add a page or delete an existing one, or edit one using the **Edit** option. To add a page to the website, select the **Add Menu Item** option available at the bottom of the menu, as shown in the following screenshot of the pop-up window:

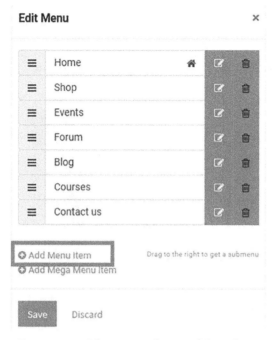

Figure 9.33 – Edit Menu web page of the website

On choosing to **Add Menu Item**, you will get a pop-up window as seen in the next figure:

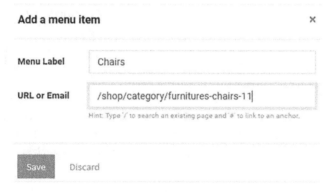

Figure 9.34 – The pop-up window to add a menu item

Here, you need to provide a **Menu Label** for the page and provide a **URL or Email** for the visitors to be directed to. Moreover, in providing the URL, you should start the description with the / character and further assignations should also follow this, as depicted in *Figure 9.32*. Furthermore, upon providing the details, you should **Save** the menu.

Back in **Edit Menu**, you will be able to view the new web page that has been created, just as depicted in the following screenshot:

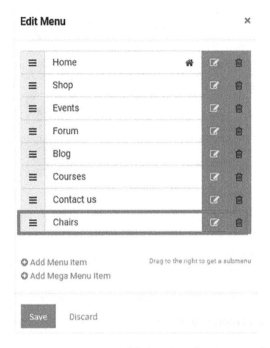

Figure 9.35 – A new page added to the Edit Menu window

Here, you have the provision to delete the web page as described before. If you choose to **Save** the menu, the web page will be added to the website.

Once the menu is added, you can select it to view all the products described under the respective product category. Moreover, the products will be listed out just as described by you, and visitors can select each one to be purchased. Furthermore, there is a search bar and filtering option available to extract the required product from the long list available. You can also choose the price list based on the user levels and categorization by brand:

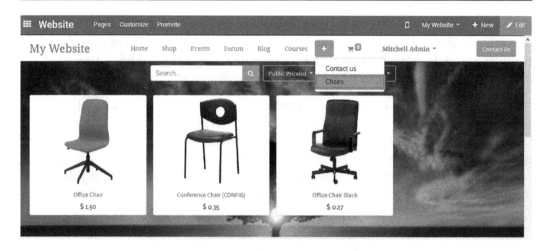

Figure 9.36 – The added product category window of chairs

The option to add a web page of the product categories will allow visitors of your website to directly view the products rather than filtering them out or searching for them from the main products menu where all the products of your platform are listed out, which might have a huge number of items.

Let's now move on to understanding the various payment acquirers that can be configured in e-commerce operations run on the Odoo platform.

Configuring different payment acquirers for customer payments

Payment acquirers are the tools available in operations that were introduced into the business world upon the establishment of digital and card payments. Today there are various payment acquirers available in the world – both local and international ones. The Odoo platform understands the need for and vitalness of payment acquirers in the operations of e-commerce platforms and provides you with the provisions to define the ones available.

By default, the Odoo platform will have auto-installed payment acquirers that are mostly supported internationally. However, if you need to integrate the platform with one of the payment acquirers that operates locally, you should initially define it and configure it to be operated based on the standards put forward by the service-providing company.

The payment acquirers defined on the Odoo platform can be accessed from the **Accounting** module to configure it to be operational on an e-commerce web page. In the **Accounting** module, select the **Payment Acquirers** menu from the **Configuration** tab and you will be shown the menu where the various payment acquirers are described.

In the menu, you can see the various payments acquirers, both enabled ones and the disabled ones, in operation on the Odoo platform. You can directly click the **Activate** and **Deactivate** options to trigger the respective operations. Some of the internationally available payment acquirers that are defined on the Odoo platform are **Odoo Payments by Adyen**, **Paypal**, **Ingenico**, **Authorize.Net**, **Adyen**, **Allpay**, and many more, as shown in the following screenshot of the **Payment Acquirers** menu:

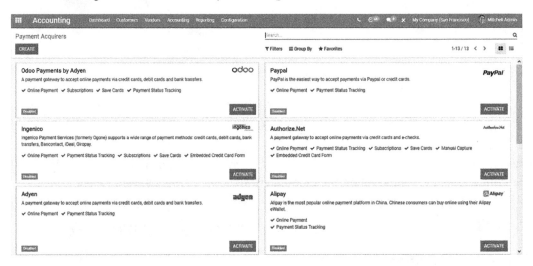

Figure 9.37 – Payment Acquirers menu

Let's have a look at the configurational aspects of the payment acquirers and the various options to define them on the platform. The name of the payment acquirer is defined, the **State** of the operation can be modified, and the **Company** can be specified if functioning on an Odoo platform with multiple companies in operation. Similarly, the **Website** on which the respective payment acquirer is in operation can be defined as depicted in the following screenshot:

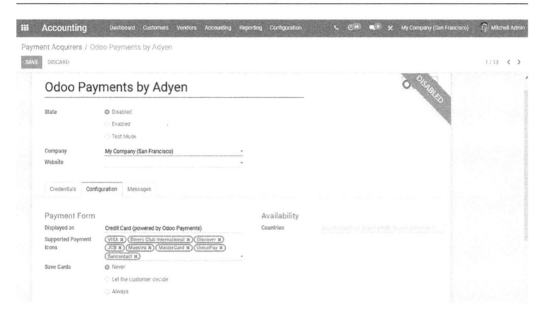

Figure 9.38 – Description menu of Odoo Payments by Adyen

Further, under the **Configuration** tab of the payment acquirer description, **Payment Form** details such as **Displayed as** and **Supported Payment Icons** can be added. The **Availability** aspect of **Countries** can be also defined under the **Configuration** tab. In addition, you can define **Credentials** as well as **Fees** involved with the respective payment acquirer in the respective windows. Not only this but a message to be displayed when the customer uses the respective payment acquirer can be defined in the **Messages** tab.

The **Save Cards** option can be assigned as **Never, Let the customer decide**, or **Always**. Additionally, the **Payment Flow** details and configurations can also be set in the window:

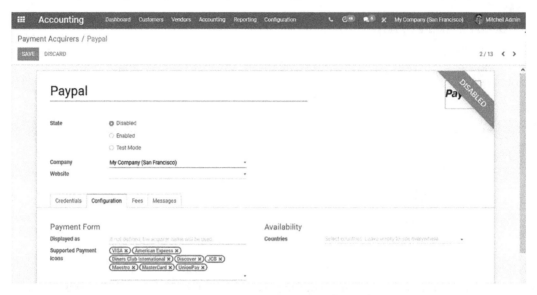

Figure 9.39 – Description menu of PayPal

Up until now, we have been discussing the existing payment acquirers available on the Odoo platform. However, there will be instances where you should define the payment acquirer yourself as part of the localization aspect of the Odoo platform to be functional in your region or country. To do so, you can create new payment acquirers on the platform by selecting the **Create** option available in the **Payment Acquirers** menu. Upon opting to create a new payment acquirer, you will be shown the following window to describe the various aspects of it.

Initially, you can provide a name, assign the **State**, **Company**, and the **Website** it functions on. Furthermore, the various aspects of **Configuration** can also be defined, such as **Displayed as**, **Supported Payment Icons**, **Enable QR Codes** or disable this, along with the **Communication** aspects of the payment acquirer. Moreover, the **Availability** of **Countries** can also be defined:

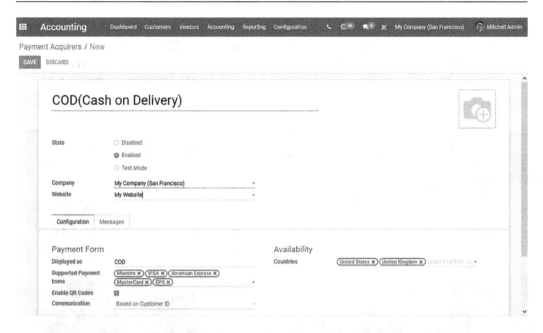

Figure 9.40 – Payment Acquirers creation window

Additionally, you can configure custom-made **Messages** to be displayed to the customer upon choosing to pay with the respective payment acquirer, which can be further modified as per the requirements:

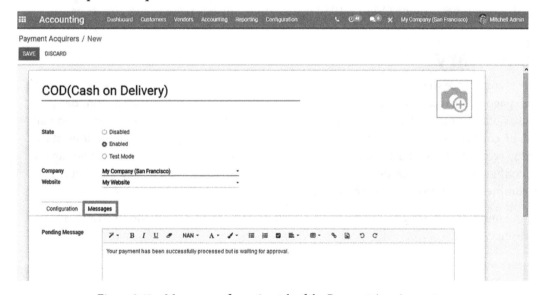

Figure 9.41 – Message configuration tab of the Payment Acquirers setup

Once the payment acquirer is saved and made active, the customer or the website visitor can choose a product to purchase. The visitor then reaches the **Billing & Shipping** menu as shown in the following screenshot. It will list out the payment acquirers you created to choose from in the list available, along with the supported payment icons to choose from:

Figure 9.42 – The created payment acquirers available in the billing menu of the e-commerce web page

The payment acquirers option in Odoo is a reliable method of accepting payment from various platforms available both internationally and locally. Moreover, it contributes to the localization aspects of the Odoo ERP in operation in various regions and countries.

Summary

The e-commerce website of your company is defined on the company website as a separate web page in operation that can be efficiently managed with the Odoo website builder tools and available functions. Now, you have an understanding of how an e-commerce web page can be configured, designed, and developed using the various operational functions of the Odoo website builder. Furthermore, you will have an understanding of and will be capable of configuring the various options available in the Odoo website builder to support the e-commerce operations of your company.

In the next chapter, we will be focusing on aspects of creating a discussion forum for website visitors.

Questions

1. How can we add a product to an e-commerce website run in Odoo?

2. Can we use any number of product category web pages for a website?

3. Can we configure a new product category from the website?

Further reading

- *Working with Odoo* by Greg Moss, Packt Publishing
- *Learn Odoo* by Greg Moss, Packt Publishing

10
A Discussion Forum for Your Clients

In the previous chapter, we were discussing the various aspects of go-live for your e-commerce website and running it efficiently with the help of advanced tools available in the Odoo website builder. We discussed adding a product to an e-commerce page and providing styling and appearance configuration for the contents using the block tools available in the Odoo website builder. We also discussed how to create a separate page for product categories and to configure various payment acquirers to the e-commerce platform with the help of the accounting module of Odoo ERP.

In this chapter, we will be discussing how to configure the forums on a company website and its various aspects, such as the following:

- Creating a forum
- Designing a forum
- Managing forum questions and answers

By the end of this chapter, you will be able to create new forums on your website and manage them efficiently using the dedicated tools available on the Odoo platform.

Technical requirements

Initially, you will require a system with Odoo ERP installed on it or accessible from it with a moderate functioning and processing speed. Furthermore, you will require an understanding of the operation of Odoo ERP and the configurations and functioning aspects of the Odoo website builder, which you might have if you have been following the book chronologically based on the chapter numbers. In addition, a knowledge of forums and their uses will be an added advantage.

Creating a forum

You might have heard about online forums and forum meetings that take place publicly as well as within an organization. So what exactly is a forum? Forums can be defined as a meeting that can happen virtually or physically or can be a platform where ideas and discussions on a particular topic are shared. Moreover, a forum can be a common ground to exchange information on topics and to provide your views and support on a topic. A forum can also act as a *Q&A* arena for web page visitors to ask questions on a particular topic and either the other visitors themselves or website owners can post answers. The *Q&A* can be based on a particular product or service of the company or based on a common aspect or topic of public interest.

In this era of the digitalized world, digital forums have more of an advantage over meetings conducted in person, thanks to the ability for people from all parts of the world to be able to participate in an online forum with the help of various virtual meeting applications, chat rooms, social media platforms, and much more.

Considering a company website, a forum page within the website will be of greater advantage in attaining public opinion on an issue. This could be based on a product for which the company would like to know about the required features that the public needs, along with the pricing as well as the allocated aspects. A public forum can be created where the customer can point out their views and the features that they would like in a product. Furthermore, forums can be discussion platforms where information is exchanged and views shared on a particular topic can be defined on the e-learning aspects.

On the Odoo platform, forum posting and forums are regulated by the **Website** module, where you can define and create forums and postings based on it, and further manage them. The forums created are depicted on the website as a separate page where the candidates, as well as the users, can post their views on the topic as well as answering the questions put forward.

To create forum postings, you should initially install the **Forum** module from the **Apps** menu of the Odoo platform. The **Forum** module is a supporting module of the website application of Odoo and you can install it by choosing the available **Install** option, as shown next:

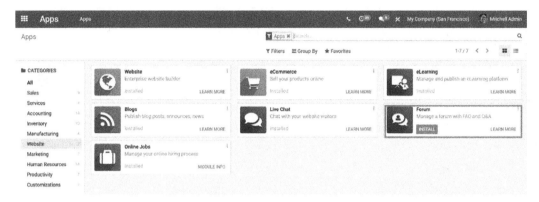

Figure 10.1 – Website based modules in the Apps menu

As the module has been installed, you will be shown the **Forum** tab in the **Configuration** menu of the website module, which can be chosen to view the **Forums** window as depicted in *Figure 10.2*. Here, all the forums described on your website are depicted – both the published as well as the unpublished ones. Moreover, if the forum is for a specific website run by the company, the website name is listed, otherwise, if it's for all the websites of the platform, it's left blank. Furthermore, the number of **Posts**, **Views**, **Answers**, and **Favorites** are shown.

Forum Name	Website	# Posts	# Views	# Answers	# Favorites
⊹ Help		6	23	5	0
⊹ Basics of Gardening		6	23	5	0
⊹ Trees, Wood and Gardens		6	23	5	0
⊹ Are fries good?	My Website	6	23	5	0

Figure 10.2 – Forums tab in the Website module

To create a new forum, you can select the **CREATE** option from the **Forums** tab which will direct you to the forum creation window as depicted in *Figure 10.3*. Here you can provide a **Forum Name**, assign the **Mode** of the forum – it can either be a **Questions** based one or a **Discussion** based, allocate the designated **Website**, and add an image. Furthermore, under the **Options** tab, you can configure **Order and Visibility** options such as **Default Sort** and **Privacy**. The **eLearning** settings where the **Course** has been described can also be mentioned.

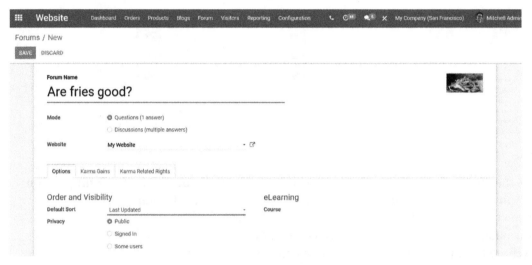

Figure 10.3 – Forum creation window

The points and rewards for the candidates and the public who take part in the forums are provided as karma points, which can be exchanged to purchase various products and services of the company. The points description is provided on the **Karma Gains** tab as depicted in *Figure 10.4*. Here, the points for operations such as **Asking a question**, **Question upvoted**, **Question downvoted**, **Answer upvoted**, **Answer downvoted**, **Accepting an answer**, **Answer accepted**, and **Answer flagged** can be described too.

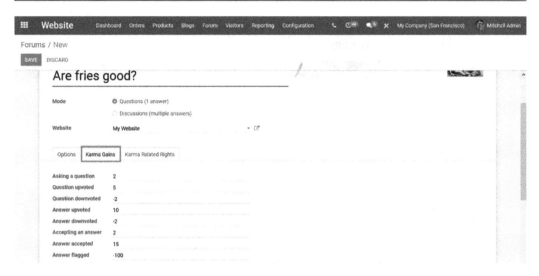

Figure 10.4 – Karma Gains menu in forum creation

In addition, on the **Karma Related Rights** tab, you can define the points for the various aspects of operation under the forum. Here you'll find, the points for operations such as **Ask question, Answer question, Upvote, Downvote, Edit own post, Edit all post, Close own post, Close all post, Delete own posts, Delete all posts, Nofollow links, Accept an answer on own questions, Accept an answer to all questions, Editor Features: Image and links, Comment own posts, Comment all posts, Convert own answers to comments and vice versa, Unlink own comments, Unlink all comments, Ask questions without validation, Flag a post as offensive, Moderate posts,** and **Change question tags.**

Once all the options are configured, you should save the respective forum. Therefore, it will be operated based on the configuration you have provided:

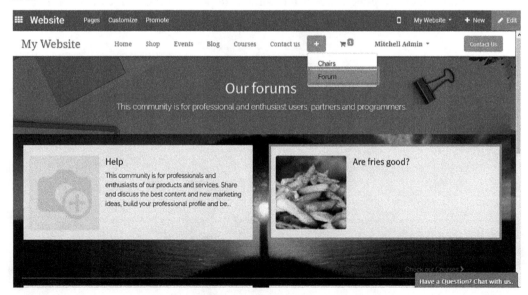

Figure 10.5 – Karma-related right menu in forum creation

Now, as forums are saved and configured, as visitors log into the website, they will be shown the **Forum** menu option. Upon clicking it, they will be directed to the forums page just as depicted in the following screenshot. Here, all the forums that you have described on the platform will be depicted and the user can select each one to participate in it:

Figure 10.6 – Forum menu on the website

In the respective forum window, all the posts of the forum will be described along with the allocated details that you have assigned. Furthermore, you, along with all authorized users, will be able to create a new post in the forum by selecting the available **New Post** option. In the window, you have the options to view **My Posts**, **Favourites**, **Followed Posts**, and **Followed Tags**. In addition, you have various Moderation Tools that will help you to opt for **To Validate** and **Flagged** posts:

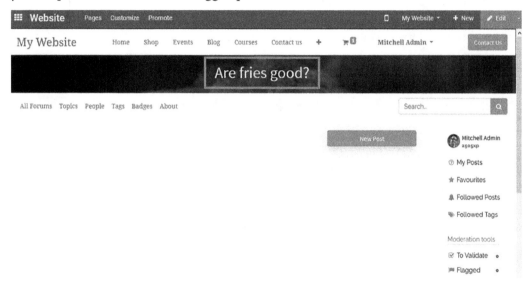

Figure 10.7 – Created forum on the website

Once the forum has been created and has been operated on, it will be depicted in the **Forums** menu, as shown in the following screenshot. Here, all the basic information of the forum, as shown before, will be depicted:

	Forum Name	Website	# Posts	# Views	# Answers	# Favorites
	Help		6	23	5	0
	Basics of Gardening		6	23	5	0
	Trees, Wood and Gardens		6	23	5	0
	Are fries good?	My Website	6	23	5	0

Figure 10.8 – The respective forum in the menu

The forums will be a common platform for discussions on a company topic and can be based on the products and services of a company too. Until now, we have been discussing the creation and management aspects of forums. Let's now move on to understand the design elements of them in the next section.

Designing a forum

The Odoo website builder, just as in all other web page design operations, provides you with ample tools to configure the appearance and contrast of a web page as per your description. Considering the **Forums** page, the use of various available design and configuration elements will add up to the appearance and styling aspects of the web page. Like every web page operation in Odoo website builder, we have an **Edit** menu here as well. This menu can be accessed from the wedge by choosing the **Edit** option, available in the top right-hand corner, as depicted in the following screenshot:

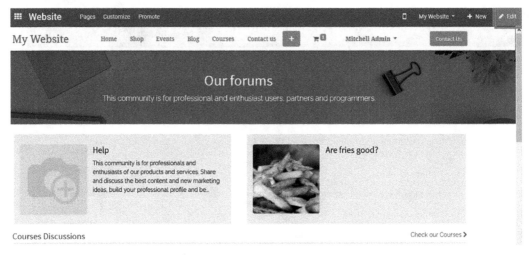

Figure 10.9 – The Edit option on the forum web page

In the editing window, just as is available on all other web pages, you will have the **BLOCKS** tools described to choose from, along with the **STYLE** and configuration **OPTIONS** available to modify the appearance of the default structure of the block and the web page. In the editing window, you will have a variety of block tools to choose from, which will help you to configure the various contents as well as the web page elements that you can describe on the **Forums** page. Furthermore, you can choose any blocks and drag and drop them in the desired location on the **Forums** page.

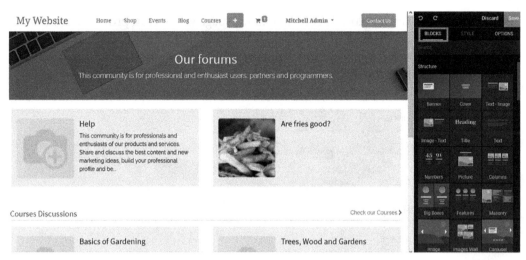

Figure 10.10 – BLOCKS menu for forum web page editing

The **Discussion Group** block of the Odoo website builder will provide users with direct access to a forum channel that has been created by you via email communication. The **Discussion Group** block is available under the **Dynamic Content** block tools of the Odoo website builder, which can be dragged and dropped on the web page as depicted in the following screenshot:

Figure 10.11 – Discussion Group block on the web page

With the **Discussion Group** block tool in action, web page visitors can provide their email address to subscribe to a discussion forum where the communications occur using the email services. Moreover, if you want to configure the available **Subscribe** button, double-click on it and you will be shown the **Link to** menu as shown in the following screenshot. Here, the button can be configured for aspects such as **Type**, **Size**, and **Style**:

Figure 10.12 – The Subscribe button editing window

The **Teams** block is one of the block operations that you can use on the **Forum** page as we will be able to describe to the people which company representatives will be contributing to the forum other than the general public who have access to it. The team members can be defined in the block along with their avatar and a brief description of them, as depicted in the following screenshot. Moreover, there are various style and configuration options available in the **STYLE** and **OPTIONS** menus that will help you to configure the appearance of the **Teams** block as per your requirements.

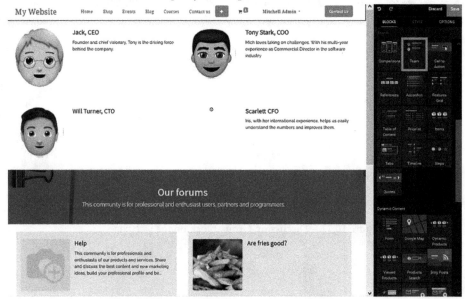

Figure 10.13 – Teams block on the web page

Another block tool is the **Dynamic Products** block, which will help you to describe the various products that can be depicted on the **Forums** page. Moreover, it will provide access to the visitors to directly go to the e-commerce web page of the website to purchase a product.

Figure 10.14 – Product comparison block on the web page

In the Odoo website builder, numerous block tools will be provided to you with dedicated design elements to describe your web page content. The block tools are classified into four categories:

- **Structure** block
- **Features** block
- **Dynamic Content** block
- **Inner Content** block

Within these classifications, there are various block tools to choose from.

You should've read *Chapter 3, Introduction to Blocks – Structure Blocks*, to understand **Structure** blocks. Further, to learn about the **Features** block, you would've *Chapter 4, Design Using Features Blocks*. In addition, *Chapter 5, Designing a Website using Dynamic Content*, describes the **Dynamic Content** blocks of Odoo. Additionally, to get an understanding of the **Inner Content** block, you can refer to *Chapter 6, Inner Content Block Tools*. Revisit those chapters for a quick refresher. Up until now, we have been discussing the design aspects of the **Forums** web page using block tools. Let's now move on to the next section, where we will be focusing on the management aspect of forum questions and answers.

Managing forum questions and answers

You might now be able to define the forums operations on the Odoo platform. However, creating and managing forums is one of the easier operation tasks, but the importance of the management of questions and answers in operations is great.

The **Forum Posts** window can be accessed from the **Posts** menu in the **Forum** tab of the website modules dashboard as seen in *Figure 10.15*. Here, all the forum postings will be listed and further details on them, such as the **Forum** on which it was posted, **Views** of the post, **Answers** given in the post, **Favorite** posts, and the **Status** of the post are shown.

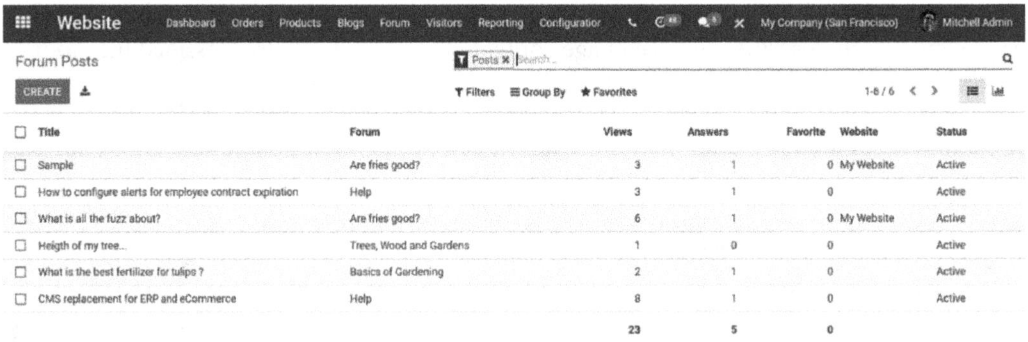
Website
Dashboard Orders Products Blogs Forum Visitors Reporting Configuration
My Company (San Francisco)
Mitchell Admin
Forum Posts
Posts
CREATE
Filters Group By Favorites
1-6 / 6
Title Forum Views Answers Favorite Website Status
Sample Are fries good? 3 1 0 My Website Active
How to configure alerts for employee contract expiration Help 3 1 0 Active
What is all the fuzz about? Are fries good? 6 1 0 My Website Active
Heigth of my tree... Trees, Wood and Gardens 1 0 0 Active
What is the best fertilizer for tulips ? Basics of Gardening 2 1 0 Active
CMS replacement for ERP and eCommerce Help 8 1 0 Active
23 5 0

Figure 10.15 – The Forum Posts menu in the Website module

Furthermore, you can create a new post for any forum that has been described on the Odoo platform by selecting the available **Create** option, upon which you will be shown the following post creation window. Here, you can provide a name for the post, on which **Forum** it should be posted, the **Website** it has been assigned to, and **Tags** can be allocated, along with the **Status** and **Reason** for the respective status.

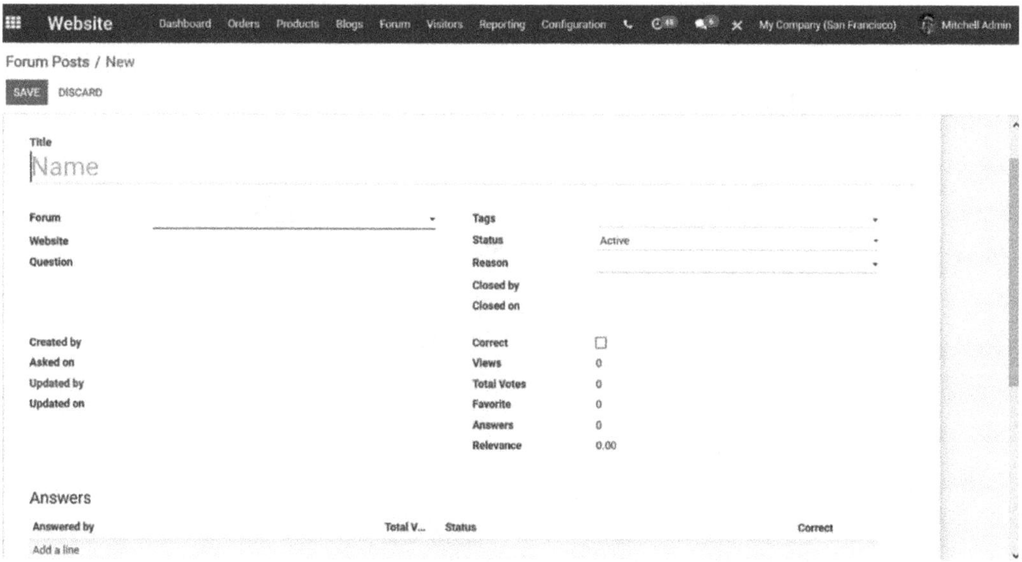

Figure 10.16 – Forum Posts creation window

Further, if you want to add an answer to the post, you can select the **Add a line** option in the **Answers** section. Here, upon choosing to add an answer, you will be shown the **Create Post Answers** window, as shown in the following screenshot.

Here, the answer **Title** can be provided, and the **Forum** should be posted on the **Website** it has been assigned to. **Tags** can be allocated along with the **Status** and **Reason** for the respective status. Finally, choose the **SAVE & CLOSE** option if you have finished adding answers. Otherwise, to add a new answer, choose the **SAVE & NEW** option.

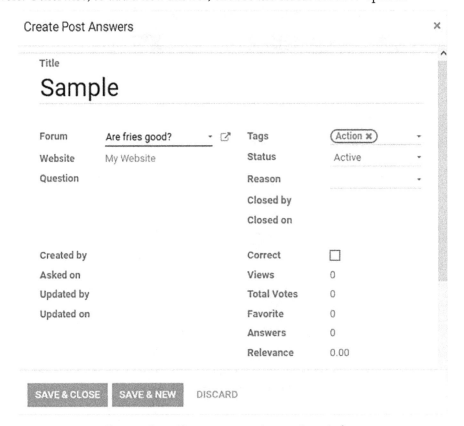

Figure 10.17 – Forum post answer creation window

Now when a visitor logs onto the website, all the added answers and the sub-post under the post of a forum will be depicted in the respective window of the web page as depicted in the following screenshot. Here, the answer on the forum will be depicted as a post along with the sub-post for the respective answer:

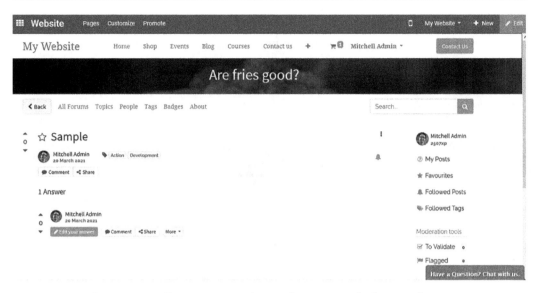

Figure 10.18 – Forum post questions and answers on the forum web page

Furthermore, you can choose to edit a post by selecting the available **Edit your answer** option, which will direct you to an **Edit Answer** window in the frontend itself, as depicted in the following screenshot. The answer editing option is provided to all users who have authorized access. After finishing editing, you can select the **Save Changes** option. Here are the breadcrumbs on the web page that will be shown for you to navigate back to the post:

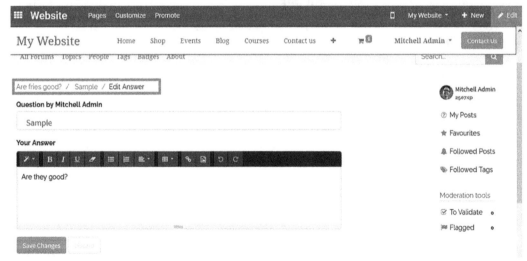

Figure 10.19 – Forum post Edit Answer window

In addition, you can add a new post to the forum directly from the website by selecting the **New Post** option on the respective forum page as shown in the following screenshot. This will allow you to directly create a new post that will be shown in the forum and also in the backend of the platform.

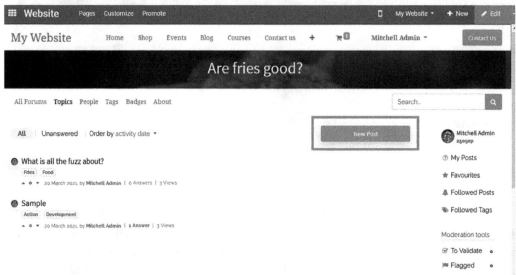

Figure 10.20 – Forum New Post creation option

In the post creation window, you should initially provide a **Title** for it, describe the post in the **Description** window, and allocate **Tags** for the post. Moreover, you can directly create new **Tags** from the window itself. Finally, to post the description, select the **Post Your Question** option or, otherwise, the **Discard** option.

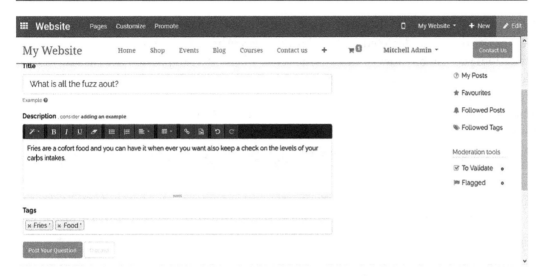

Figure 10.21 – Forum post creation window

The posted question will be depicted in the forum and you can select it and perform operations on it such as **Edit**, **Delete**, **Flag**, or **Close**, as depicted in the following screenshot. Furthermore, all the configuration settings here on the post will be auto-depicted in the backend of the platform:

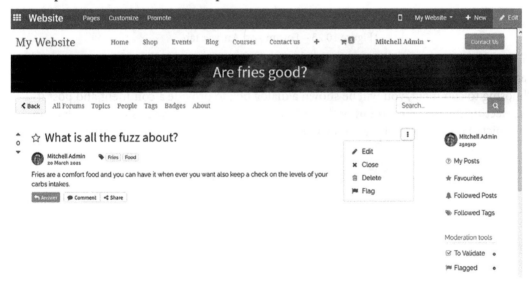

Figure 10.22 – Forum post edit options

To answer a question described in the post, you can select the respective post and select the **Answer** option, as depicted in the following screenshot. Upon choosing to answer a question, you will be shown a description block where you can provide the answer. Moreover, you will have various content configuration options for the style, appearance, alignment, an option to attach media and links, and many more to choose from to provide uniqueness and customization for it. Finally, choose the **Post Answer** option and the answer you provide will be posted.

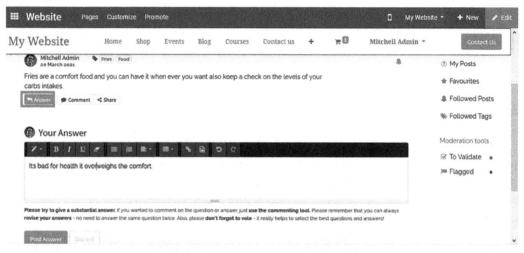

Figure 10.23 – Forum post Answer tab

Furthermore, to comment on a post, a question, or an answer under a post, you can choose the **Comment** option available under each post and description. Upon choosing to comment, you will be shown a dialog box to provide a comment and post it by selecting the **Post Comment** option. Otherwise, **Discard** it. The posted comment will be depicted under the post and you will have the option to delete or modify the comment in the future.

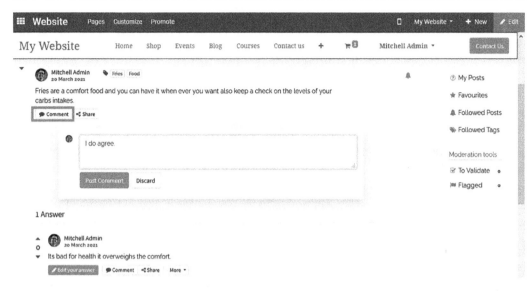

Figure 10.24 – Forum post Comment tab

Additionally, if you want to share a post or a question under the post, you can opt for the **Share** option available under the respective fields. Upon choosing to share, you will be able to share post contents on social media platforms such as Facebook, Twitter, or LinkedIn if your respective social media account has been configured.

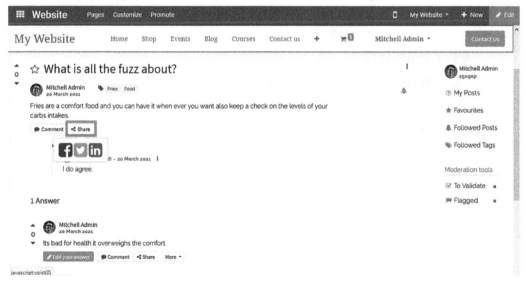

Figure 10.25 – Forum post Share option

With the various options that have been described before, you will be able to configure the operations of forum post questions and answers as per the requirement you and your website have to offer. In addition, with direct integration with the backend of the platform, all website forum operations will be in alignment and fully controlled by you.

Summary

In this chapter, we focused on forum posting as well as the management of forums on the Odoo platform. Moreover, we discussed how to create a forum, manage it, design a forum page to suit your requirements, and finally understood how postings can be managed for both questions as well as answers. As of now, you will have a clear understanding of how the forum operations of Odoo perform and how to configure it for the desired operations.

In the next chapter, we will be focusing on website tracking aspects and the related option available in the Odoo website builder, and how to configure it effectively.

Questions

1. How are the participants of forums rewarded in Odoo?

2. Can we define these karma points and post each forum with customized points?

3. Is there an option to share the posts and the comments on a forum?

Further reading

- *Working with Odoo* by Greg Moss, Packt Publishing
- *Learn Odoo* by Greg Moss, Packt Publishing

11
Tracking Your Website with Odoo

In the last chapter, we discussed **forums** and how they can be created for various purposes, including discussions and Q&A sessions. We then moved on to discuss the design element of the Forums web page, using the various block tools available in the Odoo website builder. In this chapter, we will be focusing on website tracking, and how this can be done with the various tools available in the Odoo website builder. Since Odoo has a distinct menu for tracking operations, this chapter will describe those aspects. Moreover, these dedicated tools will be explained in detail with the help of examples.

The topics covered in this chapter are as follows:

- Introduction to website tracking
- Tracking your website visitors in Odoo

By the end of this chapter, you will be able to track the web page visitors of your website, using the tools and options available in the Odoo website builder and the website module.

Technical requirements

Web page tracking in Odoo can only be done with the website module in operation. So, initially, you should install Odoo in a system that has a moderate operating speed and a reliable processor. Furthermore, install the website module for the applications menu of the Odoo platform. In addition, you should have a basic knowledge of website tracking and an understanding of how it works.

Website tracking – an intuitive feature

Website tracking is vital for every organization as it provides an insight into how the website is being utilized by web page visitors. Moreover, as it is the product of the company, they should have a full understanding of the operations involved in it. Therefore, website tracking has become an essential part of website operations. Moreover, the ability to track visitors will be vital as this will provide you with an insight into visitors' activities and whether they are involved in any malpractices while using it, and to identify the visitor if any legal obligations occur, based on the website.

With the website tracking functionality enabled in operations, you will be able to view those pages that visitors have skipped on and the time spent by each visitor on the respective web pages, images, and videos on the website. Moreover, the clicks on the various menu items, along with the options available, can also be tracked and analyzed with the web analysis tools. Moreover, an analysis of the web-based activities of visitors will allow you to have an understanding of the web page content that interests them, and accordingly, you can modify the content and the appearance of the web pages. Furthermore, with dedicated tools for Search Engine Optimization, you can provide keywords in relation to the contents described to ensure that your website shows up when visitors make relevant searches.

Another aspect of website tracking is to obtain leads on potential customers and clients for a particular product or service. The product and services of your company will be described on the company website or the e-commerce platform. An analysis of visitors searching for similar products, as well as those spending time on the product or service pages, can be undertaken. After retrieving the visitors' details, you will be able to generate marketing emails and online advertisements for them. Moreover, you can concentrate on providing pop-up advertisements or notifications to visitors based on their searches. With this method, you can follow up the web page visits and pursue them with activities such as calls, meetings, promotional messages, and many more to turn them into potential customers.

Let's discuss some of the advantages of website tracking:

- You will be able to analyze web page visitor activities and, accordingly, generate promotional and marketing advertisements targeting the audience.

- An analysis of visitors will allow you to understand the needs of customers and visitors, allowing you to modify web page content accordingly.

- You will be able to spy on visitors and track their use of the website for legal reasons.

- The web page tracking review will provide you with details regarding the usability of the website and those aspects where a web page has an error in terms of not functioning as per its optimal design and description.

These are some of the advantages of using website tracking operations, which will provide you with ample data on how the marketing aspects of your company should be directed. Let's now move on to understanding how website tracking operations are conducted in Odoo in the next section.

Tracking your website visitors in Odoo

Website tracking operations in Odoo are well equipped with functional options, which will allow you to understand visitors to the web page, and also provide you with an analysis of visitor activity on your website. Moreover, operations are interlinked with the Google Analytics tool, which is one of the prominent website tracking tools available today. The Google Analytics tool is a product offered by Google that will help you run website tracking and perform analyses on the website traffic.

Google Analytics can be used to understand the functional behavior of the visitor, website content, and device operation, and allows you to generate analyses of website operations. Today, most website managers and SEO teams all across the world rely on Google Analytics to execute their best website projects and generate website-based customer leads and business opportunities.

To run your website tracking operations in Google Analytics, perform the following steps:

1. You should initially create the website projects in Google and get an **API key** for the operation. For that, you can go directly to the **Google Cloud Platform**. Or from Odoo, you can be directed to the respective window by selecting the **Create a Google Project and Get a Key** option available in the **Slides** sections, under the **Features** pane of the **Settings** tab in the **Website** module, as depicted in the following screenshot:

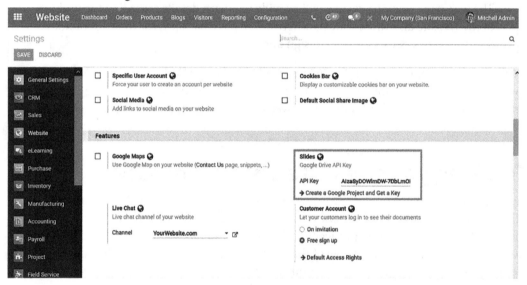

Figure 11.1 – Slides option in the settings tab of the Website module

2. Upon selecting **Create a Google Project**, you will be presented with the **Google Cloud Platform** window, which you can configure initially by registering your application to Google Drive API. Here, you should read all the terms and conditions before agreeing to the **Terms of Service** option. Select **Country of residence** and choose the **Agree and continue** option available, as shown in the following screenshot:

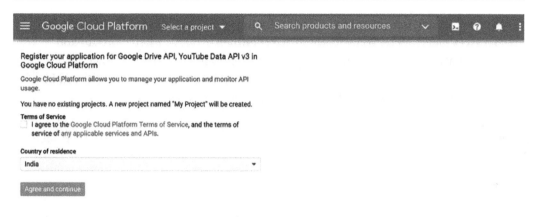

Figure 11.2 – Google Drive API registration window

3. Once the APIs are enabled, you will come across the window as shown in *Figure 11.3*. In addition, you have the provisional option, **Go to credentials**, to configure the credential of the account as shown here:

Figure 11.3 – Window depicting the successful enabling of APIs

4. Upon choosing the **Go to credentials** option, you will be directed to the **APIs & Services** menu. Here, you will see the **Credentials** window, where you can add credentials to your project by configuring the options available. Moreover, you can configure the options such as **Which API are you using?**, **Where will you be calling the API from?**, and **What data will you be accessing?** as follows:

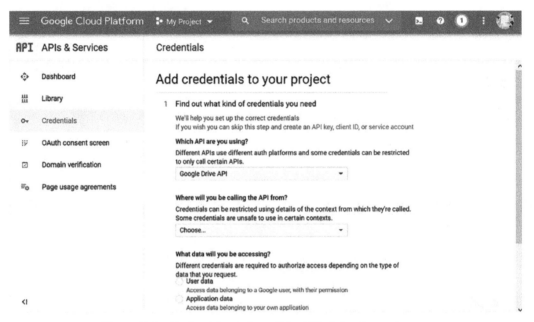

Figure 11.4 – Credentials window of the APIs & Services web page

5. Once the credentials have been configured, you can see the **Traffic**, **Errors**, and **Median latency** options, along with many more in the **Dashboard** tab of the **API & Services** menu. These are shown in the form of graphs to provide you with an understanding of website operations, as shown here:

Figure 11.5 – Dashboard window of the APIs & Services web page

If you want to create a new project to be in operation, you can select the **My Project** option available in the **Dashboard** menu of the **API & Services** window, and you will see the following pop-up window:

Figure 11.6 – Projects pop-up window of the APIs & Services web page

Here, all the projects in operation on the platform will be available, and you can create a new one by selecting the **NEW PROJECT** option.

In the **New Project** creation window, you should provide the project name, the organization in which it functions, and the website's location of operation. After providing the details, you can either opt for the **CREATE** option to create new projects, or the **CANCEL** option to discard the new project's entry. You can see the **New Project** creation window here:

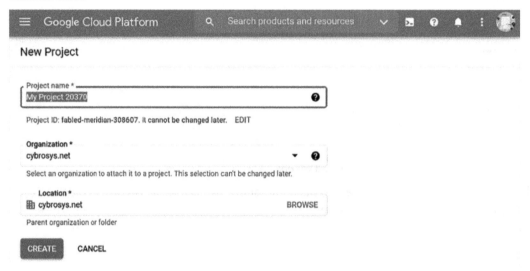

Figure 11.7 – New Project creation window of the APIs & Services web page

Additionally, under the Google Cloud Platform, there will be a number of APIs in operation. You can choose all APIs that should be in operation for the respective projects from the **API Library** window, as shown in the following screenshot:

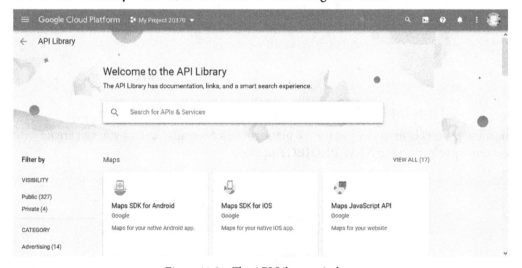

Figure 11.8 – The API Library window

Moreover, the APIs are available for all Google applications, including **Maps**, **Google Workspace**, **Advertising**, **YouTube**, and many more, which can be enabled and chosen based on the project's operations and requirements.

You can choose any one of the APIs available, and you will be directed to the respective window, where it is described. In the following screenshot, we have chosen **Google Drive API**, which can be enabled by selecting the **ENABLE** option available:

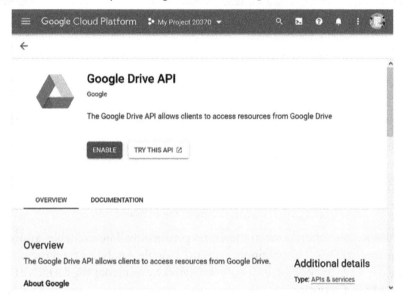

Figure 11.9 – Google Drive API window of the Google Cloud Platform

In addition, you can select the **TRY THIS API** option to have a demo of it. Furthermore, the **OVERVIEW** and **DOCUMENTATION** options of the API are available in the lower part of the web page to provide you with further information pertaining to it.

Now will have an understanding of how to configure the various Google APIs for the Odoo platform and its functionality in analyzing the web page visitors. In the next section, we will be discussing the tracking aspects based on the Google Analytics tool, which can be configured in Odoo.

Website tracking using Google Analytics in Odoo

Google Analytics, as described previously, is one of the most prominent website tracking tools available today, and it can easily be configured to be operational with the Odoo platform. From the **Website**, module, select the **Analytics** option from the dashboard and you will be presented with the menu as in the following screenshot:

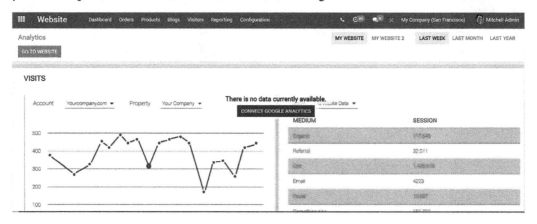

Figure 11.10 – Analytics menu of the website module

Moreover, all the website operations in the Odoo platform will be available, and you can choose the one that you want to configure with the Google Analytics tracking operations. Here, initially, you have to configure Google Analytics by selecting the **CONNECT GOOGLE ANALYTICS** option.

Now, you will see the corresponding menu pop up, as shown in the following screenshot:

Figure 11.11 – Connect Google Analytics menu popup

Here, you can provide details for the **Your Tracking ID** and **Your Client ID** fields and click **SAVE**. Additionally, you can choose from **How to get my Tracking ID** and **How to get my Client ID** options, which will direct you to the documentation web page available on the Odoo platform, provided you have internet access to the system.

Now that you have an understanding of how the Google Analytics tracking option can be configured, let's move on to understand the Link Tracker option, available in Odoo, in the next section.

Website tracking using the Link Tracker option

Another website tracking functionality in Odoo is the **Link Tracker** option, which can be availed to track web page operations described inside the website. To configure the **Link Tracker** option, you should initially install the **Link Tracker** module from the applications menu of the Odoo platform. You can perform the following steps to run your website tracking operations in Odoo:

1. To install Link Tracker, select the **INSTALL** option, and it will be available for operation. The following screenshot shows the **Apps** window of the Odoo platform, where the category is set as **Website**, and all the filter components are removed to get the **Link Tracker** module:

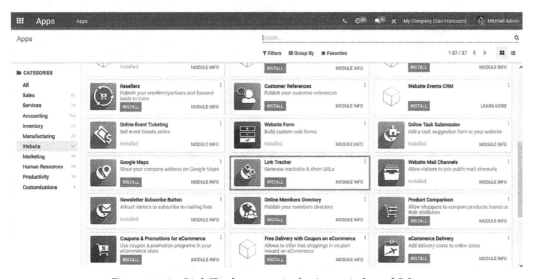

Figure 11.12 – Link Tracker menu in the Apps window of Odoo

2. Once the application is installed, the **Link Tracker** option can be viewed on the website of the platform. You can log in to the website and choose the **Promote** option available on the dashboard. You will be able to view the **Link Tracker** option as shown in the following screenshot:

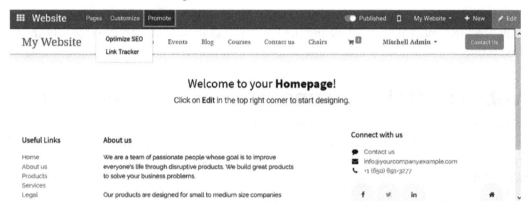

Figure 11.13 – The Promote option on the website

3. In the **Link Tracker** menu, you can create a new link tracking operation by providing the URL, assign the campaign, the medium of operation, and the source in which the web page is described. Finally, select the **Get tracked link** option. In addition, the **Your tracked links** menu will display all the links that you are tracking, as shown in the following screenshot:

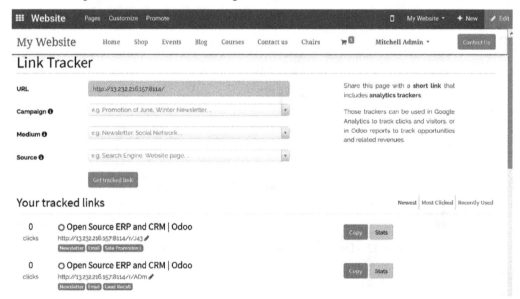

Figure 11.14 – Link Tracker web page on the website

4. Moreover, the tracked links can be filtered with options such as, such as **Newest**, **Most Clicked**, and **Recently Used**. The following screenshot shows the filtered results using the **Most Clicked** filter option:

Figure 11.15 – Most Clicked filter on the Link Tracker web page

5. You can choose any of the default filtering options to view the tracking option of a respective web page of your choosing. Additionally, you can select the respective link by clicking on the **Statistics** option, which will show the status of the link tracker. In the respective **Link Tracker** menu, the number of **clicks**, along with the tracking description and the tags, will be shown as follows:

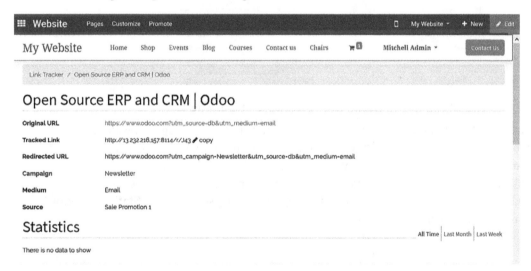

Figure 11.16 – Web page link tracker for your website

The **Link Tracker** functionality of the Odoo website allows you to track the operations of the web pages individually. You will be able to understand visitor activity, and the statistical data on it will allow you to plan the contents of the web page. In addition, the Link Tracker operations are also available in the backend of the platform. For that, you can initially search for the Caps application and, upon entering it, you will be presented with the window depicted in the following screenshot:

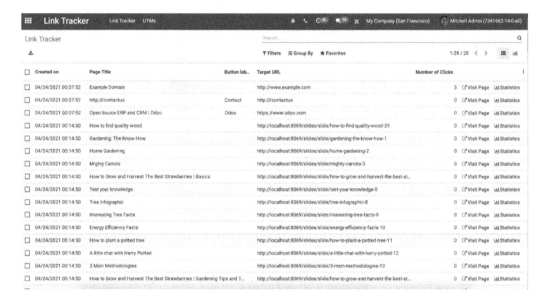

Figure 11.17 – Searching for the link tracker

Here, the **Created** date and time will be described, along with **Page Title** and **Button label**. Furthermore, you will have information pertaining to **Target URL**, as well as the **Number of Clicks** count. Moreover, you can visit the respective web page directly by clicking on the available **Visit Page** option. Additionally, the statistics on each of the web pages can be viewed as described in *Figure 11.16* by selecting the **Statistics** option available next to each **Visit Page** link.

Summary

In this chapter, we focused on the website tracking aspects of the Odoo website builder and how it is helpful in the website management operations of a company. We discussed how to configure the Google APIs, got an understanding of the Google Analytics tool in the Odoo platform, and had an insight into the Link Tracker functionality available in Odoo. Moreover, with all these covered, you now have an understanding of the tracking operations and will be able to function with ease in Odoo.

In the next chapter, we will be focusing on the aspects of drafting a Contact us page for your website and configuring it to meet your requirements.

Questions

1. How many new projects can we create in the Google API?
2. Can we track multiple links in the Link Tracker option of the Odoo website?
3. Are we able to view the number of clicks on a link using Odoo?

12
Drafting a Contact Page

In the previous chapter, we focused on the website tracking aspects of operations using the Odoo website builder. We focused on the various tracking tools, such as **Google APIs** and services, **Google Analytics**, and the **Link Tracker** options available with the Odoo website builder, as well as covering the website module of the platform. Now, you should have an understanding of the website tracking aspects and how the website tracking tools available in Odoo function. In addition, you should be able to configure a web page or website tracking by referring to the previous chapter.

In this chapter, we will be focusing on drafting a contact page for a website using the Odoo website builder. The following are the concepts that we will look at in this chapter:

- Creating a new contact form
- Designing the form
- Lead generation using a contact form

By the end of this chapter, you will be able to draft a new contact form and draft a new web page with a dedicated form.

Technical requirements

You will require a system with the latest version of Odoo installed, or you should have subscription access to the platform. Furthermore, your system should be able to perform reasonably well. Basic knowledge of the Odoo platform and its website builder tool would be beneficial, along with an understanding of website form creation.

Creating a new contact form

A contact form is one of the easiest ways for website visitors or customers of a company to communicate with the website owner or the company. In a contact form, visitors provide their name and contact details; along with that, they can also post a question or a comment to the company based on their products, services, or a query. Web page forms are also one of the best ways to receive messages from visitors without any involvement of external parties and other platforms.

Another aspect of the contact page is the functionality of opportunity generation, which can be done by using the lead generation functionality available on the company website. With the Odoo website builder tool, you can create a new web page for the operations of the contact form on a website. Choose the **New** option on the website and select the **Page** option, as shown in the following figure:

Figure 12.1 – Menu to add a new element to a website

In the **New Page** creation window, as shown in *Figure 12.2*, you can provide the page title and then click **Continue**:

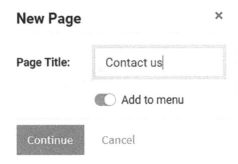

Figure 12.2 – The pop-up window to provide the web page name

Upon providing the page title and clicking **Continue**, you will be able to view the web page option on the dashboard. Upon choosing it, you will be presented with a blank page. To bring in the contact form, select the **Form** block from the **Dynamic Content** block tools available in the **Edit** menu of the web page. The block will provide you with a default structure of content, as shown in the following figure, that can be edited:

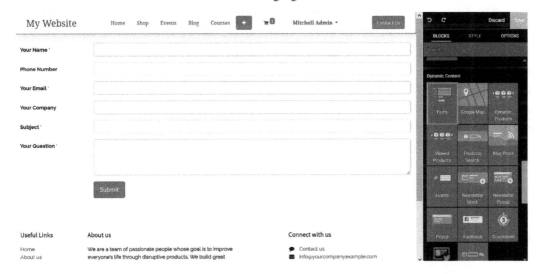

Figure 12.3 – Form block for the Contact us web page

The default content, such as the questions and the fields, can be modified from the **Fields** menu as depicted in the following figure. Here, the **Type, Height, Input Placeholder, Label Name,** and **Label Position** fields, and the options, such as **Required, Hidden,** and **Shown on mobile,** can be enabled or disabled:

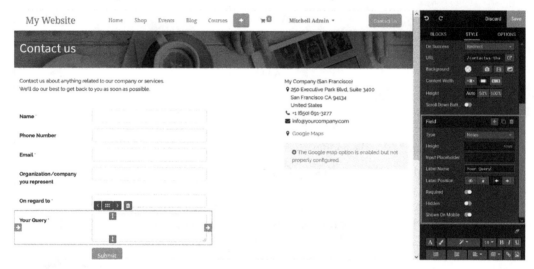

Figure 12.4 – Field editing menu of the Contact us web page

As the contact page details and the fields have been provided now, let's move on to the **Actions** aspect of the **Form** menu, depicted in *Figure 12.5* under the **STYLE** menu. The action of an operation after the visitor fills in the contact page details can be configured under the **Actions** tab. The action can be set as **Apply for a Job, Create a Customer, Create an Opportunity, Create a Task,** or **Send an Email.** These actions will all perform various tasks according to what is selected. If you select **Send an Email,** you need to provide an email address for **Recipient Email.**

Fields can be marked as **Required** or **Optional** under the **Marked Fields** tab. The label width of the fields can also be provided. After the completion of the form, the visitor should hit the **Submit** button as depicted in the following figure. The action for the **Submit** button can be configured in the **On Success** tab. Here, either the user can be redirected to a different web page or you can opt for the **Show message** option. The URL for the redirecting operations can be configured under the **URL** tab.

Furthermore, the background for the **Contact us** page can be set as a photo, video, or pattern. The content width and height can also be configured. You have the option to enable or disable the **Scroll Down** button. Also, the **Submit** button can be configured using various configuration options such as **Button Position** and **Show On Mobile**:

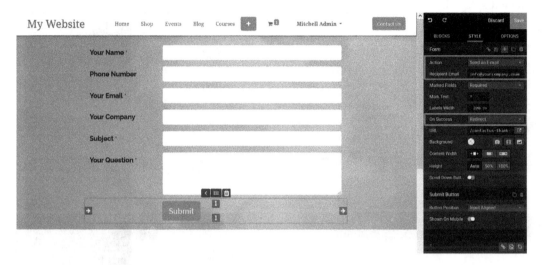

Figure 12.5 – STYLE editing window for the form block

You should now understand how a contact form can be configured for a company website using the Odoo website builder. Now, let's move on to the next section, where we will learn how various leads and opportunities can be generated using content forms on a website.

Lead generation using a contact form

Lead generation is considered one of the vital aspects of company operations. Companies should have multiple ways of lead generation, in order to maximize business opportunities. The Odoo ERP accounts for the need for lead generation and management and provides you with multiple tools for it. One such tool in Odoo is the lead generation aspect of the contact form, which can be configured under the **STYLE** menu available in the editing menu shown in *Figure 12.6*.

Here the action is configured as **Create an Opportunity**, meaning now you will be able to choose a sales team and a salesperson, provide the marked fields, and set all the other configuration options. Furthermore, the **On Success** option can be set to **Show Message** so that after the visitor fills in the details and selects the **Submit** button, a success message will be displayed:

Figure 12.6 – Contact us action configuration menu

The configured success message block can be edited and provided with a message as shown in *Figure 12.7*:

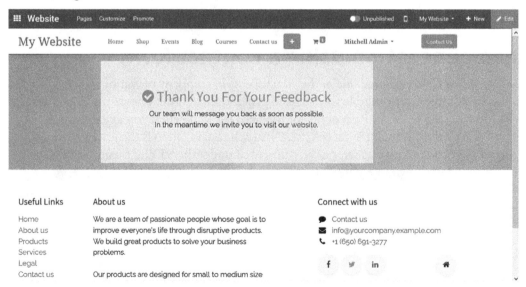

Figure 12.7 – Success message displayed to visitors after submitting the Contact us form

Once the message has been received in the company server, the Odoo platform will recognize it and automatically turn the visitor information into lead information for the CRM module. The lead information will be what was provided on the **Contact us** page. Moreover, the leads will be shown in the pipeline as shown in the following figure:

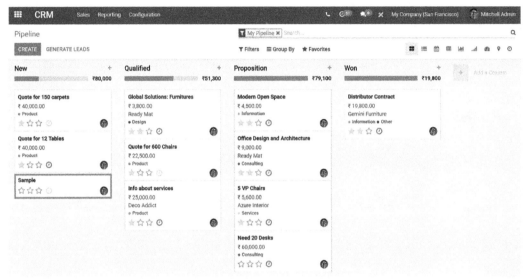

Figure 12.8 – Leads in the pipeline menu of the CRM module

Once you enter the respective lead, you will be presented with the content that the visitor provided on the **Contact us** page. Operations on leads such as **NEW QUOTATION**, **MARK WON**, **MARK LOST**, and **ENRICH** can be set. All the opportunity enrichment and business attaining functions can be performed using the lead pipeline window in the CRM module. These functions bring more business opportunities and financial gains to the organization:

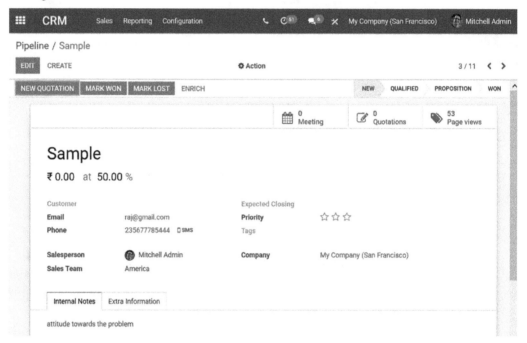

Figure 12.9 – Lead in the CRM module

A lead generated through the **Contact us** page of a website can be followed up with various activities that can be scheduled with the customer to pursue the lead and convert it into an opportunity. So far, we have been discussing lead creation from the **Contact us** page. Now let's move on to the other actions for which the **Contact us** page can be configured in the next section.

Other actions for the Contact us page

Now it should be evident that the **Contact us** page can be configured in many different ways according to your requirements. The Odoo platform itself has put forward certain default actions that we will cover now. Moreover, you can create and configure various other customizable actions for the backend as per your requirements. The following figure depicts the various actions that can be configured for the form page of a website:

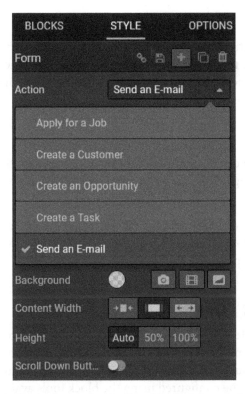

Figure 12.10 – Actions under the form block of the website builder

Applying for a job

The **Apply for a Job** action can be chosen to display at the end of the **Contact us** page. Upon choosing **Apply for a Job**, the visitor who fills out the form will have their application for the respective job submitted, and the application for the job posting will be depicted in the **Recruitment** module of the Odoo platform.

Creating a customer

The **Create a Customer** action can also be used, whereby Odoo will automatically create a customer in the **Contacts** module based on the information from the form.

Creating a ticket

In Odoo, you can run online as well as offline events with the help of the dedicated **Events** module of the platform. The **Create a Ticket** action for the **Contact us** page will allow you to create and generate tickets based on the requirements of the website visitor. Moreover, if tickets are to be generated after payment, that too can be configured.

Creating a task

With the **Create a Task** action enabled for the **Contact us** page, web page visitors can ask you to perform a task. The requested task can be created and assigned from the **Projects** module of the platform. These tasks may be manufacturing or repair operations, for instance.

Sending an email

The **Send an Email** action can be configured for operations where the customer needs an acknowledgment message. This can be an auto-generated acknowledgment message configured via Odoo.

These are the default types of action tools available; others can be configured based on your requirements from the backend of the platform. Let's now move on to understanding the design elements of the platform.

Designing the form

The form page can be made to be one of the best and most attractive pages of a website while providing various functional options for visitors to move within the website and providing them with various marketing and promotional information. The design aspects of the **Contact us** page can be configured using the block tools available in the Odoo website builder. Moreover, you have numerous block tools under **Structure**, **Feature**, **Dynamic Content**, and **Inner Content**.

Chapter 3, Introduction to Blocks – Structure Blocks, will provide you with all the details on the configuration and design aspects of **Structure** blocks. Furthermore, to find out about the **Features** block tool, read *Chapter 4, Design Using Features Blocks*. Additionally, *Chapter 5, Designing a Website using Dynamic Content*, will describe the **Dynamic Content** blocks tools of Odoo, along with the configuration and design aspects. In addition, to get an understanding of **Inner Content** blocks and their configuration and design aspects, you can refer to *Chapter 6, Inner Content Block Tools*.

Here, to show the operation, we will insert the **Banner** block under the **Structure** block classification into the **Contact us** web page. You can describe the contents along with the design elements on the block and configure the **Contact us** button, which will direct visitors to the form:

Figure 12.11 – Banner block on the Contact us web page

Another block tool to discuss here is the **Team** block; using this, you can show the team members of your company. The **Team** block can be selected from the **Features** section and then dragged and dropped into the web page location. Furthermore, the editing options, as well as other configuration options, will be available in the **STYLE** and **OPTIONS** menus:

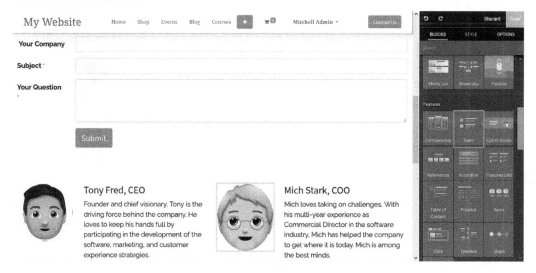

Figure 12.12 – Team block for the Contact us web page

Now you should have an understanding of the block operations for the **Contact us** web page. Now let's move on to editing the fields of web page. Before describing a field, we should provide the type of the field, which can be set as **Selection**, **Text**, **Long Text**, **Email**, **Telephone**, **URL**, **Number**, **Decimal Number**, **Author**, **CC**, **Channels**, **Country**, **Region**, **Sales Teams**, **Salesperson**, and more.

The options need to be configured based on your requirements; you can add new options as well as removing unnecessary fields. The label name along with any additional configurable options, which can be enabled and disabled, can also be provided as shown in the following figure:

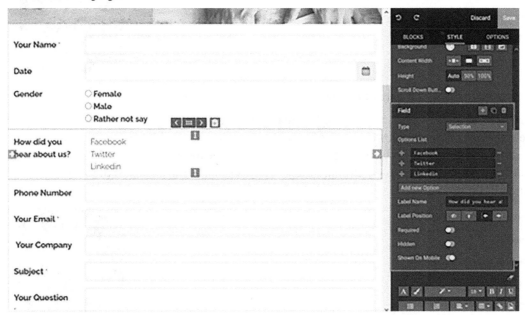

Figure 12.13 – Field editing menu of the form block for the Contact us page

Once all the configurable and style editing functions have been provided and described, you need to save what you have done:

Figure 12.14 – Resultant Contact us web page after the field and block editing

The web page design and styling aspects we have covered will provide you with what you need to bring in visitors and turn them into possible business opportunities.

Summary

The **Contact us** page of the website is one of the key elements to facilitate communication between web page visitors and the organization. Moreover, it also acts as one of the best lead and opportunity generation tools that a company can have, enabling the company to follow up on more leads and opportunities in Odoo to turn into business proposals. In this chapter, we discussed how to create a **Contact us** web page, add various design elements, and configure it for application-specific operations. We also discussed the use of block tools to change the page's appearance.

Now you will be able to design a contact page for your website using the Odoo website builder and configure the various aspects of it based on your requirements. In addition, with the help of the block tools, you will be able to configure the appearance fully.

In the next chapter, we will be focusing on how the live chat functionality can be configured.

Questions

1. Are all the block tools of Odoo available on the **Contact us** web page?

2. Which block tool will bring in the contact form to the web page?

3. How do you change a visitor message into a lead?

Further reading

*   *Working with Odoo, Greg Moss, Packt Publishing*

*   *Learn Odoo, Greg Moss, Packt Publishing*

13
Communicating with Live Chat

In the previous chapter, we were focusing on the contact form page for the company website that can be used to generate leads and business opportunities for the company. Moreover, we learned how the **Contact us** page can be created and managed, how **Fields** are provided, and how design elements are provided and depicted to be in operation. Furthermore, you will now have an understanding of how the contact us web page can be created using the Odoo website builder tool.

In this chapter, we will learn about the aspects of configuring a Live Chat tool, including the following:

- Adding live chat to the Odoo website
- Managing incoming chat
- Assigning chats to your employee

By the end of this chapter, you will be able to configure and manage the live chat tool available in the Odoo website builder and configure it to aspects of your company operations.

Technical requirements

A system with good processing speed and functionality in terms of operation, and with the Odoo platform installed, is a prerequisite. Moreover, an understanding of the aspect of functioning with Odoo, its website building aspect, and basic knowledge of the live chat option in the website will be ideal as regards performing the operation described in this chapter.

Adding live chat to the Odoo website

The live chat functionality of a website will act as a communication tool for visitors to the company or the service provider. Moreover, it's a good way of promoting the company's products and services to the customer in a customized way. The live chat functionality of the website has the same impact as a customer communicating with a salesperson in a retail store, with the only difference being that the communication happens virtually and in the e-commerce platform of the company's website.

Today, almost every website has a live chat functionality, and this is one of the company's prominent communication and marketing tools, all thanks to developments in communication and internet facilities, which have their roots in thesis development. Although having a live chat tool won't work miracles for your company, aspects of effective management will. The Odoo platform has a dedicated Live Chat tool that integrates directly with the company website and will provide you with effective operations management.

The Live Chat module of the Odoo platform can be installed from the applications menu of the software by following these steps:

1. In the **Apps** menu, select **Website** under **CATEGORIES**.

2. Under the classification, the **Live Chat** module will be shown to you. You can install it by selecting the **INSTALL** option, as seen in the following screenshot:

Figure 13.1 – Apps menu with the Website category filter

3. Once the module has been installed, you can select the respective module from the home dashboard of the platform and you will be directed inside the dashboard of the **Live Chat** module. In the menu, all the channels that are active in operation will be depicted and you can have an initial understanding of the channel, such as the number of operators and sessions. In addition, there is a **LEAVE** option:

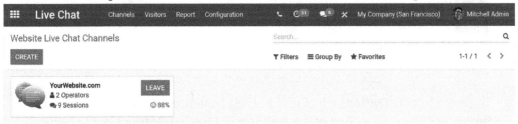

Figure 13.2 – Live Chat module dashboard

4. To create a new channel, you can select the **CREATE** option available in the menu, in the **Website Live Chat Channels** creation window. You can initially provide a name, and on the **Operators** tab, you can select **ADD** to add operators to the website:

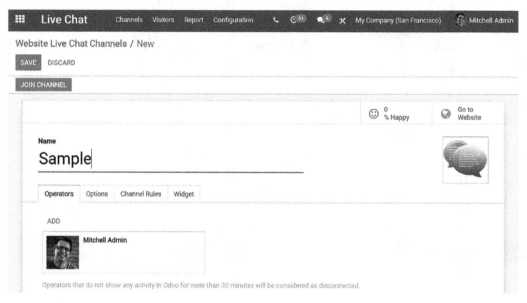

Figure 13.3 – Live Chat channel creation window

5. You can select the **Options** tab of the channel creation menu where the **Livechat Button** options, such as **Text of the Button** and **Livechat Button Color**, can be configured. In addition, the **Livechat Window** options, such as **Welcome Message**, **Chat Input Placeholder**, and **Channel Header Color**, can be defined:

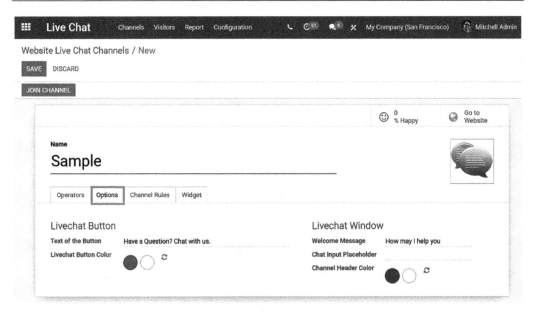

Figure 13.4 – Options menu in the Live Chat creation menu

6. Additionally, **Channel Rules** can be described in the respective menu by selecting the **Add a line** option. These rules can be custom designed using the various default options available in the menu, which can be defined for a given URL action and the country it is functional in:

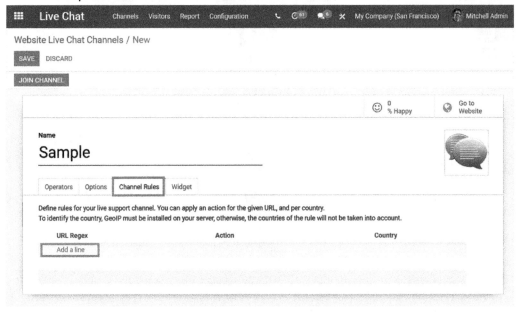

Figure 13.5 – Channel Rules menu in the Live Chat creation menu

In the **Create Rules** menu, you will see an **Add a line** option, which can be used to configure **Action** as **Display the button**, **Auto popup**, or **Hide the button**. Additionally, **URL Regex** and **Country** can also be configured:

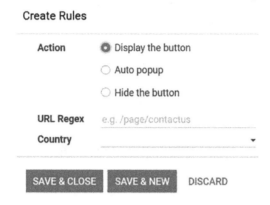

Figure 13.6 – Channel rules creation menu

Furthermore, the widget for the Live Chat function in the website can also be configured in the **Widget** menu. However, the widget configuration can only be executed after the channel has been saved, as shown in the following screenshot, with the alert message to save the channel:

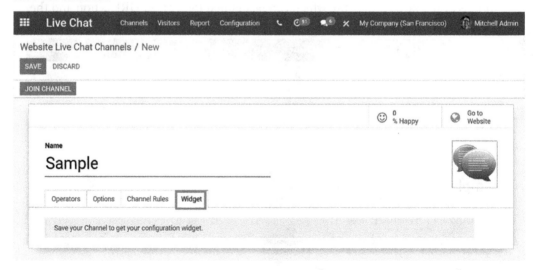

Figure 13.7 – Widget menu in the Live Chat creation menu

Once the channel has been saved, you can view the widget configuration options, which can be done using the coding aspects of the website. Once the widget has been configured as per your specifications, the Live Chat channel can be viewed with the same widget in the display from the website:

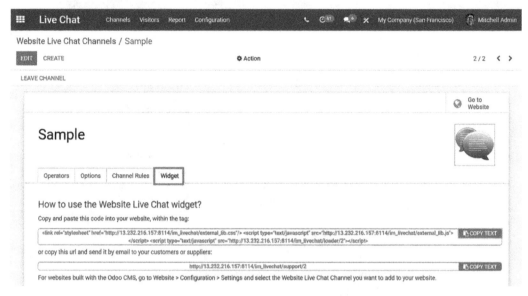

Figure 13.8 – Widget editing menu in the Live Chat creation menu

The Live Chat menu will describe the channels in operation as depicted in the following screenshot, where all the channels have been listed. Furthermore, these channels can be filtered, grouped, or sorted to retrieve the data required:

Figure 13.9 – Live Chat module with the created channel

Once the Live Chat channel has been configured, you can select the **Go to website** option available and you will be directed to the channel page of the company website, as depicted in the following screenshot:

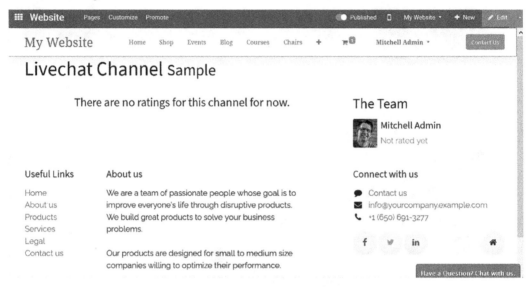

Figure 13.10 – Respective Live Chat channel in the website

Let's now move on to understand how we can create and manage multiple channels for multiple websites using the Odoo website builder and the Live Chat module of the Odoo platform in the next section.

Multiple channels for multiple websites

You can configure multiple channels for the multiple websites described in the Odoo platform. As you may be aware, you can run multiple websites for the company with Odoo, which can be done from the same website module of the platform. In case you need to run multiple channels for the Live Chat operations on each website, this can also be done by initially defining multiple channels in the Live Chat module and then configuring it to the respective websites from the website module of the platform.

Furthermore, if you need to configure a single Live Chat channel to be operational for all the websites that have been defined in the platform, you should initially create a Live Chat channel in the module and configure it for each website in the website module of the platform.

You will now have an understanding of the Live Chat tool available in the Odoo website builder and will know how to add it to your company website. Now, let's move on to understanding how to manage incoming chat using the Odoo website builder.

Managing incoming chat

As mentioned earlier, the Live Chat operations of your company will not become fully efficient simply by adding a tool to it. Rather, you will need a complete management solution that can provide you with the configurable option to manage the Live Chat operation of the website. For that, to be done in Odoo initially, you will need to configure the Live Chat tool to the company website.

Moreover, as you can define multiple channels in operations in the Odoo Live Chat function, however, you can use only one in operation at a time, which needs to be configured to the website. To configure the channel, select the website module, go to the **Settings** menu, and, on the **Features** tab, you have an option to configure **Live Chat**. Choose the channel from the drop-down menu. Here, all the channels that have been created will be listed.

Furthermore, you have an external link option relating to each channel, which can be chosen to be directed to the respective channel:

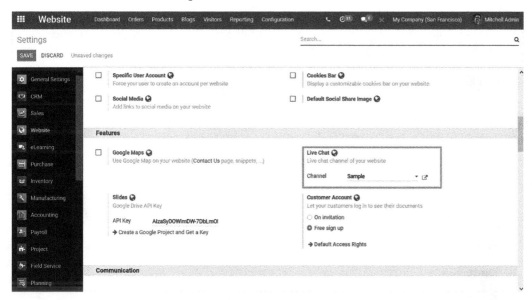

Figure 13.11 – Live Chat channel assignation option

Now, as the Live Chat functionality for the website has been configured, visitors can log on to the website to chat with your company on a variety of issues. Moreover, the Live Chat tool will be visible from any page of the company's website. A visitor can simply click on the widget and choose to chat with the company's employees directly, as shown in the following screenshot:

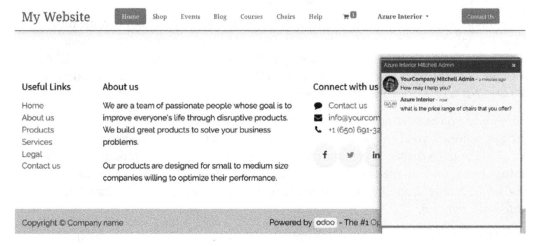

Figure 13.12 – Live Chat widget on the website

Once a visitor has sent a message, it will be displayed to you under the respective Live Chat channel. You can choose the respective channel and then select the **Sessions** menu from the dashboard, as depicted in the following screenshot:

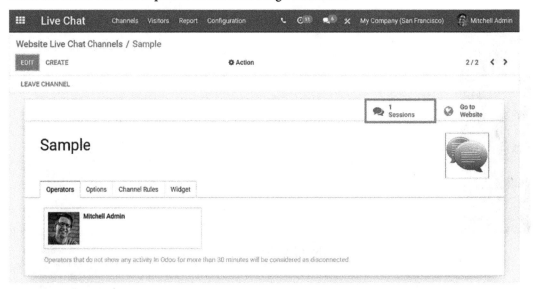

Figure 13.13 – Respective Live Chat description menu in the Live Chat module

In the **Sessions** menu, all the sessions on the respective Live Chat channel will be depicted with **Session Date** and time, **Attendees**, **Messages**, and **Rating**, if provided. Moreover, you can select the respective session to have a more in-depth understanding of the operations:

Figure 13.14 – Sessions menu of the Live Chat channel

In the respective sessions menu, you will obtain information on **Attendees**, **Session Date**, and time, as well as the **History** of the chat described, along with the time and date of the operation. However, you cannot perform editing operations on the sessions or reply to the messages directly from this menu:

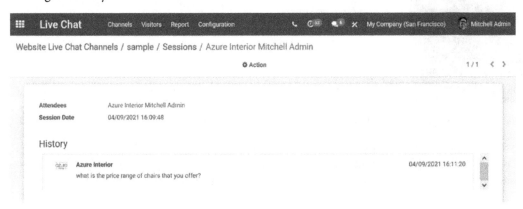

Figure 13.15 – A sessions description menu for the respective session

To reply to the chats generated from the visitor side, you can move on to the **Discuss** module and select the person from whom the chat has been generated in the form of the **LIVECHAT** menu, as depicted in the following screenshot of the **Discuss** model window. In the window, all the channels will be described and the messages from the respective visitor who has initiated a chat will be depicted under the Live Chat menu. From the list, employees can select the respective visitor and replay to the chat based on their requirements or the query:

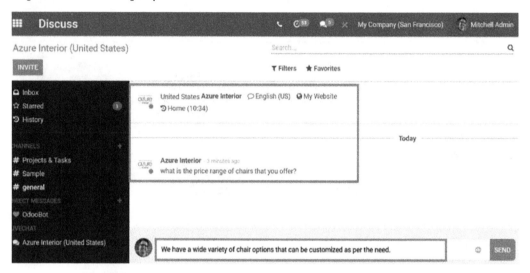

Figure 13.16 – Channel chat option for employees in the Discuss module

Messages that have been replied to will be depicted in the respective chat and you can generate one or more replies at a time. Moreover, the chat from the visitor will be auto-loaded here as employees wait in the menu to have a conversation:

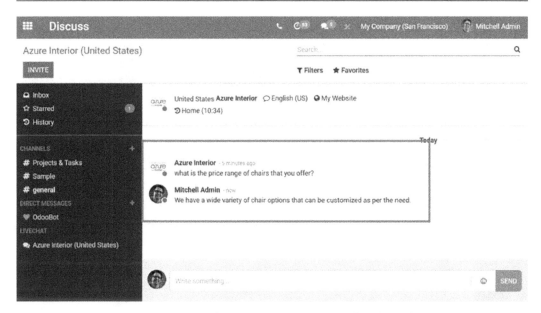

Figure 13.17 – Chat operations on the Live Chat channel

Furthermore, the discussion module of the Odoo platform is configured for an in-house discussion, as well as chat operations. This can be a beneficial tool in the company's operations. As of now, you will have an understanding of how the Live Chat operations for the company website can be configured and managed effectively. In the next section, let's move on to understanding how chats can be assigned to employees.

Assigning chats to your employee

For a website functioning on an international level, or that has visitors and customers from various parts of the world, you will receive numerous messages using the Live Chat functional tool. Moreover, company websites functioning with a wide visitor base will have messages from visitors in different languages. However, it is extremely difficult for a single employee to manage chat operations in multiple languages. Therefore, we can assign a team of employees who are well-versed in all the languages that the company and the website uses.

The employee or operators of the Live Chat channel can be added from the Live Chat module by selecting the **ADD** option, as depicted in the following screenshot:

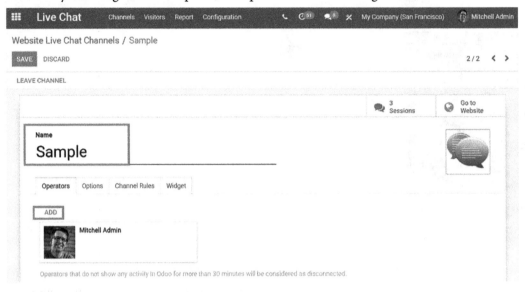

Figure 13.18 – Option to add employees to the Live Chat channel

Upon choosing **ADD** from the **Operators** menu, you will be presented with a list of employees, that have been described in the platform. Here, the name of the employee and the **Login** database, along with the **Language**, **Latest authentication**, **Company**, and **Two-factor authentication** details will be described. Moreover, there is an option to choose more than one employee by selecting the tick box available concerning each employee in the list. Furthermore, you have the option to create a new employee directly from the window, which will direct you to the web page creation menu. Once the employees are selected, you can choose the **SELECT** option to add employees to the channel:

Figure 13.19 – The Add: Operators screen for the Live Chat channel

The employees added to the channel will be described in the **Operators** menu, as depicted in the following screenshot:

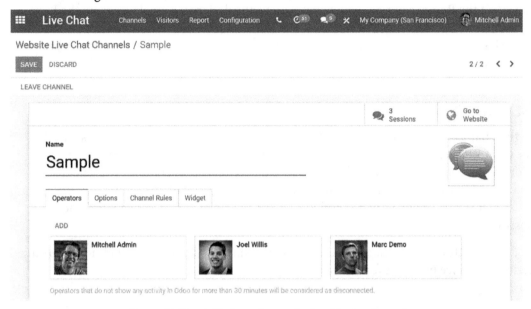

Figure 13.20 – Added employees depicted in the channel

To perform operations on the employees assigned, you can select the respective employee and a menu will be presented to you. Here, the name of the employee, their email address, and the company to whom the employee is assigned, along with their main phone number and mobile number, will be shown:

Figure 13.21 – Operator menu for the Live Chat channel

Moreover, you can edit the details provided and these will be auto changed throughout your Odoo platform. Furthermore, if you want to remove the respective employee, you can select the **Remove** option, which will remove the designated employee from the Live Chat channel. Additionally, the employee assignment operations for the respective Live Chat channels will provide you with ample opportunity to manage the operation of the Live Chat menu based on your requirements.

Summary

In this chapter, we discussed the Live Chat function of the Odoo ERP, which has paved the way for effective communication between web page visitors and company operators. Moreover, as the platform has a dedicated module to manage operations, you can create and function with numerous Live Chat channels effectively. The functioning operators for each live chat channel can be configured from the list of employees who have been described in the platform.

In the last section of this chapter, we will be concluding the book and providing you with certain reference materials for further reading on the Odoo website builder.

Questions

1. Can I use multiple employees to manage Live Chat operations in Odoo?

2. Will the Live Chat module entail any additional costs?

3. How many channels can I create and operate simultaneously?

Further reading

- *Working with Odoo*, by Greg Moss, Packt Publishing

- *Learn Odoo*, by Greg Moss, Packt Publishing

Answers

1. Yes, you can use any number of employees to manage your Live Chat channels.

2. No, there is no additional cost associated with the Live Chat module as it comes along with the basic subscription of the platform.

3. You can create any number of channels for the Live Chat operations in the Odoo platform. However, only one can function at a time. Moreover, with effective customization, you will be able to function with multiple Live Chat channels.

Conclusive overview

Earlier in this chapter, we discussed the Live Chat tools of the Odoo website builder. Moreover, we discussed managing incoming chats and assigning chats to employees. With this section, we'll conclude this book on the Odoo website builder and provide a few references that will assist you in understanding the various aspects of Odoo ERP and the tools available in it.

The Odoo website builder has now become an indispensable tool for operations regarding the functioning of the company website in Odoo. Through the tool, you will be able to create a new website, design it, and manage its operations and functions. Moreover, in this Odoo website builder book, we described website creation and how to design web pages using various blocks, such as Structure blocks, Features blocks, Dynamic Content blocks, and the Inner Content block, of the Odoo website builder.

We then moved on to the technical aspects of web page design using the HTML/CSS/JS editor. In addition, we focused on the various practical tools of the Odoo website builder, including blogs, e-commerce websites, discussion forums, website tracking aspects, drafting a contact page, and describing a Live Chat tool for the website.

Having completed this book, you will have an understanding of the Odoo website builder and will certainly be able to create, design, and manage your company websites in Odoo with ease. With the latest version of the software, Odoo 14, which is one of the most advanced and fastest Odoo ERPs ever, the Odoo website builder comes as a basic inbuilt module and further functionalities can be configured and added from the Odoo apps store.

Today, Odoo ERP is one of the prominent business management solutions available on the market, with more than 5 million users from all parts of the world. The Odoo platform has designed business operation modules that run interconnected in a unanimous way to attain complete business management for your establishments. Moreover, the Odoo website module and the website builder tools of the Odoo platform are part of an entire suite of tools available to you in Odoo that help you to run company operations efficiently.

The Odoo website module operation of the platform is integrated directly with different modules of the Odoo platform, such as the CRM module. Customer relationship management for the company operations is vital even during the operations of the company website. The e-commerce platform of the company will help you to generate leads based on website visitors and their activity.

Another module in the operation of the Odoo platform that is integrated directly with the website module is the subscription module, which will allow you to sell subscription-based products to customers. With this, you will be able to sell e-newsletters as well as magazines to customers. The e-learning module of the Odoo platform will help you to define the various learning courses for the candidates. Moreover, the module functions with the help of the company website and is defined as a separate courses page on the website. You can define various courses as well as add certification programs for the candidates to take part in them and achieve certification based on performance.

Company events or the ones conducted by the company can be defined using the events module of the Odoo platform. Here, the various events can be described, tickets for these events can be sold online to web page visitors, and payment can be obtained through the various payment means described in the platform.

References

Here are the reference materials that will provide you with further insights into the Odoo ERP and provide you with a complete understanding of it:

- *Odoo v14* book by Cybrosys Technologies: `https://www.cybrosys.com/odoo/odoo-books/odoo-book-v14/`

- Odoo user documentation from Odoo: `https://www.odoo.com/documentation/user/14.0/index.html`

Assessments

This section contains answers to all the questions presented to you at the end of each chapter.

Chapter 1: Introduction to Odoo and its Website Builder

1. The modular approach of Odoo, with a single platform and the centralized inventory and database.

2. Odoo 14 was revealed at the Odoo Experience event in 2020.

3. Website building with Odoo is easy with block tools, with a simple drag and drop functionality.

Chapter 2: The Website Builder in Action

1. Users of the Odoo platform can create a new website from the website module of the platform. In the module, enter the settings menu and you can view the option to create a new website, which can be chosen to obtain the website creation menu, where you should provide the configurable options for the website.

2. You can create a new web page from the frontend of the website by selecting the new option available from which you can select the new page option available. Moreover, from the backend of the platform, you can add it to the web pages' menu, which is accessible from the **Configuration** tab of the website module.

3. The additional themes for your website in Odoo can be obtained from the Odoo apps store and by filtering out the themes menu and should be compatible with Odoo version 14.0.

Chapter 3: Introduction to Blocks – Structure Blocks

1. There are numerous block tools in Odoo, but in general, they have been classified as three structure blocks: feature blocks, dynamic content, and inner content.

2. There are 18 different types of structure block available in the Odoo website builder tool and here are the names of a few: banner, cover, text-image, image-text, and title.

3. The block operations of Odoo will have similar types of editing options and style choices to choose from, which are detailed in the banner block sections of the chapter, and here are the names of a few: background, title, image change, and image editing tools.

Chapter 4: Design Using Features Blocks

1. In the later version of Odoo, Odoo 14, there are a total of 13 blocks that would fall into the category of the features block in the Odoo website builder.

2. The tabs block under the features block is well suited to providing navigational tools to your website visitors on the various web pages available.

3. In the Odoo website builder, the reference block coming under the category of the features block can be used by you to define the logs and images of other companies or service providers associated with your establishment.

Chapter 5: Designing Websites Using Dynamic Content

1. In total, 14 types of blocks are listed under the dynamic content block of the latest version of Odoo.

2. The dynamic blocks, such as **Facebook** and **Twitter scroller** block tools, will help you to define the social media aspects of the company on the website.

3. The Google maps can be integrated into your website using the **Google map** block available under the dynamic content block classification of the Odoo website builder.

Chapter 6: Inner Content Block Tools

1. There are a total of 13 types of inner content blocks available in the latest version of Odoo 14 in the website builder module.

2. The Text Highlight block under the inner content block classification can be used to provide highlighted content on your company's web page.

3. The Blockquote block can be the best tool to be used to define the quotes of employees and customers on the web page.

Chapter 7: Using the HTML/CSS/JavaScript Editors

1. CSS Editor deals with the styling aspects of the web page constant in the Odoo website builder.

2. HTML – HyperText Markup Language, CSS – Cascading Style Sheet, and JS – JavaScript.

3. Yes. You can insert a background image for the internet, external website, or a social media platform to the banner. However, the platform should have internet access as well as the complete URL of the image.

Chapter 8: Creating Your Own Blog Pages

1. The various blog tags can be classified and categorized under the blog categories which can be custom created by you in the blog categories menu accessible from the website module of the Odoo platform.

2. The tags can be created and allocated to the blogs under the tags menu of the website module of the Odoo platform. Here, you can create new tags, adding blogs to old as well as new tags described in the website.

3. The web page description can be provided under the SEO menu, which can be further customized as well when needed.

Chapter 9: Go Live with Your E-Commerce Website

1. A product can be added to the web page by selecting to add a product from the edit menu. Another way to add products to the web page is to add a product to the inventory and enable the listing in the e-commerce web page of the Odoo platform.

2. Yes. You can add any number of product categories as menu web pages to your e-commerce website using the Odoo website builder.

3. No. You will not be able to configure a new product category from the website. However, it can be configured from the backend of the platform where the different product categories can be created and configured, and later added to the e-commerce website.

Chapter 10: A Discussion Forum for Your Clients

1. The participants are rewarded based on the Karma points designated for each activity they perform in a forum defined by the Odoo platform.

2. Yes. You can define, customize, and provide any distinctive Karma points for the various operations in Forum.

3. Yes. You have the option to share the posts and comments directly from the forum itself to social media platforms such as Facebook, Twitter, and LinkedIn.

Chapter 11: Tracking Your Website with Odoo

1. You can create *any* number of projects in the Google API.

2. Yes. You can track multiple links using the Link tracker functionality in Odoo.

3. In the Link tracker menu, you will be able to view the number of clicks concerning each link.

Chapter 12: Drafting a Contact Page

1. Yes. All the block tools of the Odoo website builder can be availed from the editing menu of the **Contact us** web page.

2. The form block of the Odoo website builder can bring in the form for the visitor to fill in.

3. You can enable the **Create an opportunity** option available in the **Field editor** menu.

Chapter 13: Communicating with Live Chat

1. Yes. You can use any number of employees to function with your Live Chat channels.

2. No. There is no additional cost associated with the Live Chat module as it is provided with the basic platform subscription.

3. You can create any number of channels for the Live Chat operations in the Odoo platform. However, only one can function at a time. Moreover, with effective customization, you will be able to function with multiple Live Chat channels.

Packt.com

Subscribe to our online digital library for full access to over 7,000 books and videos, as well as industry leading tools to help you plan your personal development and advance your career. For more information, please visit our website.

Why subscribe?

- Spend less time learning and more time coding with practical eBooks and Videos from over 4,000 industry professionals

- Improve your learning with Skill Plans built especially for you

- Get a free eBook or video every month

- Fully searchable for easy access to vital information

- Copy and paste, print, and bookmark content

Did you know that Packt offers eBook versions of every book published, with PDF and ePub files available? You can upgrade to the eBook version at packt.com and as a print book customer, you are entitled to a discount on the eBook copy. Get in touch with us at customercare@packtpub.com for more details.

At www.packt.com, you can also read a collection of free technical articles, sign up for a range of free newsletters, and receive exclusive discounts and offers on Packt books and eBooks.

Other Book You May Enjoy

If you enjoyed this book, you may be interested in this other book by Packt:

Odoo 14 Development Cookbook

Parth Gajjar

Alexandre Fayolle

Holger Brunn

Daniel Reis

ISBN: 978-1800200319

- Learn to develop new modules and modify existing modules using the Odoo framework
- Explore key concepts of the Odoo framework to build robust business applications
- Create dynamic websites with snippets and learn to deploy an Odoo instance on the server or Odoo.sh

Packt is searching for authors like you

If you're interested in becoming an author for Packt, please visit `authors.packtpub.com` and apply today. We have worked with thousands of developers and tech professionals, just like you, to help them share their insight with the global tech community. You can make a general application, apply for a specific hot topic that we are recruiting an author for, or submit your own idea.

Leave a review - let other readers know what you think

Please share your thoughts on this book with others by leaving a review on the site that you bought it from. If you purchased the book from Amazon, please leave us an honest review on this book's Amazon page. This is vital so that other potential readers can see and use your unbiased opinion to make purchasing decisions, we can understand what our customers think about our products, and our authors can see your feedback on the title that they have worked with Packt to create. It will only take a few minutes of your time, but is valuable to other potential customers, our authors, and Packt. Thank you!

Index

W

www.ingramcontent.com/pod-product-compliance
Lightning Source LLC
Chambersburg PA
CBHW062038050326
40690CB00016B/2979